Juan M. Oyarzún

Pigment Processing

Physico-Chemical Principles

Juan M. Oyarzún: Pigment Processing

ISBN 978-3-86630-665-3

Cover: Clariant Produkte (Deutschland) GmbH

Bibliographische Information der Deutschen Bibliothek
Die Deutsche Bibliothek verzeichnet diese Publikation in der Deutschen Nationalbibliographie; detaillierte bibliographische Daten sind im Internet über http://dnb.ddb.de abrufbar.

Juan M. Oyarzún
Pigment Processing: Physico-Chemical Principles
Hanover: Vincentz Network, 2015
European Coatings LIBRARY
ISBN 3-86630-665-2
ISBN 978-3-86630-665-3

Please ask for our book catalogue
Vincentz Network, Plathnerstr. 4c, 30175 Hanover, Germany
T +49 511 9910-033, F +49 511 9910-029
books@european-coatings.com, www.european-coatings.com

Layout: Vincentz Network GmbH & Co. KG, Hanover, Germany
Druck: BWH GmbH, Hanover, Germany

ISBN 3-86630-665-2
ISBN 978-3-86630-665-3

European Coatings LIBRARY

Juan M. Oyarzún

Pigment Processing

Physico-Chemical Principles

Juan M. Oyarzún: Pigment Processing

ISBN 978-3-86630-665-3

Author

Juan M. Oyarzún was born 1932 in Chile, and studied Chemistry in Santiago. In 1960 he joined BASF, working in the technical application of pigments and dyestuffs in Ludwigshafen and Stuttgart. In 1974 he joined Farbwerke Hoechst AG (now Celanese), South Africa, as technical adviser for the application of pigment and dyestuffs. In 1978 he returned to Germany and joined Akzo Coatings in Stuttgart, where he was concerned with basic research on pigment dispersion and mill base formulations. He retired in 1997 and for a long time he acts as a free technical adviser for pigments application.

Foreword

When the author of this monograph began to concern himself nearly forty years ago with the application technology of pigment and dyestuffs, a subject that was to become his field of activity for life, one had to search long and hard for books dealing with this subject, as in those days they were scarce commodities. Fortunately, however, progress in the relevant technical literature during the last decades has gone hand in hand with the, one could almost say, galloping development in the technology of colouring materials, so that nowadays a wide selection of textbooks on production, properties, application technology and ecology of dyestuffs and pigments, in particular in the German and English languages, is at the disposal of the pigment technologist.

For this reason, and considering that pigment application technology, together with other aspects of pigment and dyestuffs technology, has already been discussed in most of the numerous, in some cases outstanding books on pigments, it would seem reasonable to ask whether there is a need for a further book dealing with this subject. As justification, the author would state his opinion that basic physical and physico-chemical features of pigment application are neglected in current technical literature and not treated in accordance with their importance in comparison with other aspects, such as manufacture, properties and uses. Even nowadays, works dealing with these important topics can practically be counted on the fingers of one hand. The present monograph thus constitutes an attempt to close this gap by offering the pigment technologist the possibility of informing himself adequately on the current state of the physico-chemical principles of pigment application technology without having to work through piles of the relevant magazines.

The author has endeavoured to explain each subject in a manner which is precise and comprehensive enough to provide the reader with a solid base of knowledge in this field of pigment technology. Those interested readers wishing to extend their knowledge on some particular aspect or other beyond the scope offered here will profit from the literature references quoted during the discussion of each topic.

This monograph thus presents a practice-related exposition of the fundamental principles of pigment processing. It is intended as a help to chemists, physicists or other graduates taking their first steps in the pigment manufacturing or processing industry to become acquainted with pigment application technology. However, in order to make this monograph accessible to the largest possible technical group, the mathematical treatment has been reduced to the necessary minimum.

In this manner, this book may also be of interest not only to colourists, technicians and laboratory assistants engaged in research, development, application or production in the pigment industry but also to those in the relevant technical colleges and research institutes.

The author cordially thanks all his former colleagues at BASF SE, Celanese (Hoechst AG) and AkzoNobel who supported him with their valuable advice, as well as the management and staff of the *Pigmente und Additive* division of Celanese (Hoechst AG) for their generous provision of electron micrographs and other pictures for illustration purposes. I would also thank Dr. Oskar J. Schmitz, former staff member of the Forschungsinstitut für Pigmente und Lacke, in Stuttgart, now retired, for his verification of the valid industrial standards.

I feel also indebted to Dr. Jochen Winkler for his contribution to this present second edition of my book concerning the ultrasonic and electroacoustic methods for determining particle size and the measurement of zeta potentials. In addition, I would like to thank Bernd Reinmüller from DIN Deutsches Institut für Normung e.V. for his contribution in verifying every DIN EN ISO norm.

Juan M. Oyarzún

Valparaíso, Chile, June 2015

Contents

Introduction **11**

1 Physical characterisation of pigments **13**
1.1 Particle morphology 13
1.2 Crystal structure 15
1.2.1 Crystal modifications: polymorphism 15
1.2.2 Isomorphism and anisotropy 16
1.2.3 X-ray crystallography 17
1.3 Particle size measurement 20
1.3.1 Particle size distribution 21
1.3.1.1 Sieve analysis 21
1.3.1.2 Determination by electron microscopy 22
1.3.1.3 Analysis by sedimentation techniques 22
1.3.1.4 Optical methods 24
1.3.1.5 Ultrasonic attenuation spectroscopy 26
1.3.1.6 Electroacoustic methods 27
1.3.1.7 The Coulter counter 27
1.3.1.8 Methods of presenting results 27
1.3.2 Specific surface area 29
1.3.2.1 The Freundlich adsorption isotherm 29
1.3.2.2 Langmuir's adsorption theory 30
1.3.2.3 The adsorption isotherm according to Brunauer, Emmett and Teller 32
1.3.2.4 Determination of the average particle size 34
1.4 Surface character 37
1.4.1 Polarity and hydrophilicity 37
1.4.2.1 Surface tension 39
1.4.2.2 Capillarity 45
1.4.3 Electrical polarisability 47
1.4.3.1 Polarisation of dielectrics in an electric field 47
1.4.3.2 Polarisability and refractive index 48
1.4.3.3 Polarisability and pigment/binder-interaction 49
1.4.4 Surface treatment 50

2	**Optical properties of pigmented systems**	**53**
2.1	Reflection, gloss and gloss haze	54
2.1.1	Gloss	54
2.1.2	Gloss haze	59
2.2	Light scattering by colloidal disperse particles	60
2.2.1	Rayleigh scattering	61
2.2.2	Mie theory	63
2.3	Absorption and scattering as factors in the tinctorial properties of pigments	69
2.3.1	Hiding power, semi-transparency and transparency	69
2.3.2	Lightening power	78
2.3.3	Tinting strength and depth of shade	79
2.3.4	Hue (or shade)	81
3	**Rheological behaviour of pigment dispersions**	**85**
3.1	Introduction	85
3.2	Newton's law of viscosity	85
3.3	Dependence of viscosity on temperature	88
3.4	Flow properties of concentrated suspensions and molecular colloids	90
3.4.1	Newtonian flow	91
3.4.2	Structural viscosity (pseudoplasticity)	92
3.4.3	Dilatancy	94
3.4.4	Plasticity	95
3.4.4.1	Bingham or linear-plastic flow	96
3.4.4.2	Non-linear plasticity (Casson viscosity)	97
3.4.4.3	Plastic-dilatant flow	100
3.4.5	Shear-time dependent flow	101
3.4.5.1	Thixotropy	101
3.4.5.2	Rheopexy	103
3.4.6	Viscoelasticity	104
3.4.6.1	Effect principles	104
3.4.6.2	Effect of molecular structure	106
3.4.7	Flow behaviour of water-thinnable binder systems	107
3.5	Rheological behaviour of pigmented systems in practice	108
3.5.1	Basic aspects	108
3.5.2	Printing inks	111
3.5.3	Paints and enamels	114

3.5.4 Plastics ... 117
3.6 Closing remarks ... 118

4 Dispersion process: physico-chemical fundamentals ... 119
4.1 Introduction ... 119
4.2 Stages of the dispersion process ... 120
4.2.1 Mechanical breakdown ... 120
4.2.1.1 Forces of interaction between the pigment particles ... 120
4.2.1.2 Types of mechanical breakdown ... 123
4.2.2.1 Energetic approach ... 125
4.2.2.2 Wetting kinetics ... 128
4.2.3 Stabilisation ... 130
4.2.3.1 Fundamental aspects ... 130
4.2.3.2 Stabilisation by steric hindrance ... 130
4.2.3.3 Electrostatic stabilisation: DLVO theory and zeta potential ... 143
4.2.3.4 Measuring Zeta potentials ... 147
4.3 Stabilisation problems with pigmented systems after dispersion ... 149
4.3.1 Flocculation ... 149
4.3.1.1 Fundamental aspects ... 149
4.3.1.2 Testing and measuring flocculation behaviour ... 151
4.3.1.3 Kinetic approach to flocculation ... 153
4.3.1.4 Coflocculation ... 155
4.3.1.5 Prevention of flocculation ... 156
4.3.2 Flooding and floating ... 156
4.3.2.1 Discussion and theoretical explanation ... 156
4.3.2.2 Testing separation phenomena. spray-pouring test and rub-out test ... 160
4.3.2.3 Methods for overcoming flooding and floating ... 161
4.3.3 Shock effects and seeding ... 164
4.3.3.1 Pigment shock ... 164
4.3.3.2 Binder shock ... 165
4.3.3.3 Solvent shock ... 166
4.3.3.4 Seeding ... 166

5 Dispersion process: methods and mathematical models for assessing the degree of dispersion ... 169
5.1 Introduction ... 169

5.2	Direct methods	169
5.2.1	Fineness-of-grind and dispersion curve	169
5.2.2	Examination with the optical microscope	173
5.3	Indirect methods	174
5.3.1	Tinting strength development	174
5.3.2	Assessment of the state of dispersion from gloss development	181
5.3.3	Increase in lightness flop	182
5.4	Concluding remarks	184
6	**Dispersion process: mill base optimisation**	**185**
6.1	General considerations	185
6.2	Description of the dispersion process by a probability function	188
6.2.1	Theoretical principles	188
6.2.2	Practical conclusions from the probability relationship for the dispersion process	192
6.3	Dispersibility of pigments	193
6.3.1	Critical approach to the definition of pigment dispersibility according to DIN ISO 18451-1	193
6.3.2	Testing and stating the dispersibility as specified in DIN EN ISO 8780 and DIN EN ISO 8781	193
6.3.3	Assessment of the dispersibility by means of a half-life value	195
6.3.4	Pragmatic definition for pigment dispersibility	197
6.4	Methods for mill base optimisation	203
6.4.1	Daniel flow-point procedure	203
6.4.2	Guggenheim factor	205
6.4.3	Efficiency of the dispersion process: new criterion for optimizing mill bases	206
6.4.4	Optimisation of several factors by means of an experimental design according to Plackett and Burman	211
6.4.5	Optimisation of the cost of the dispersion process by controlling energy input	214
	Appendix	**218**
	List of physical symbols and constants used	**218**
	List of relevant norms	**221**
	References	**225**
	Index	**234**

Introduction

The colouring materials termed pigments, whose characteristic feature is their insolubility in the application medium (DIN ISO 18451-1, DIN EN ISO 4618), are marketed as powders, press-cakes, flushed pigments or pigment preparations. The main advantage of the last three delivery forms is that the pigment processor need not disperse them further. They merely have to be stirred into the bulk medium and homogenised, thereby saving time and costs. In addition, they often enhance tinting strength and transparency, afford better colour uniformity in plastics and their handling is simple and clean.

The application behaviour of pigments is not only determined by their chemical constitution. Important physical parameters such as crystal modification, mean particle size, particle size distribution, shape, specific surface area, surface character, optical properties, rheological behaviour and dispersibility also have a decisive influence on their subsequent areas of use. For this reason, pigment manufacturers nowadays regulate the processes in such a manner that the resulting products, though being of the same chemical constitution, can exhibit different physical properties and are thus optimised from the beginning with a view to their future application field.

The theoretical principles and practical results of these fundamental physical and physico-chemical aspects of pigment application will be discussed in the present book.

1 Physical characterisation of pigments

1.1 Particle morphology

Optical and particularly electron microscope research gave rise to the development of a pigment model [1] that today forms the basis of DIN (EN ISO 18451-1) 53 206. According to this model, there are three kinds of particles present in a pigment powder. They are shown in Figure 1.1.

Crystals or primary particles
Aggregates
Agglomerates

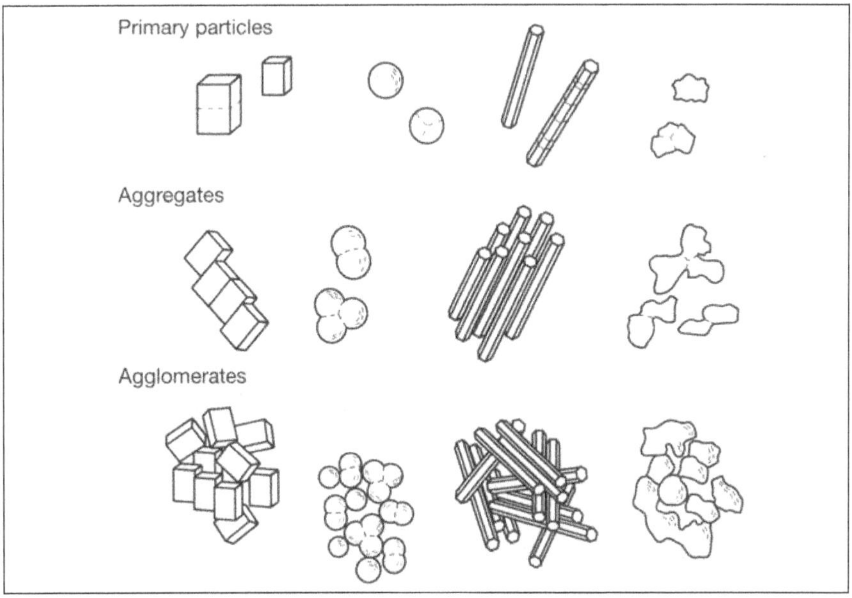

Figure 1.1: Pigment models according to EN ISO 18451-1

Juan M. Oyarzún: Pigment Processing

ISBN 978-3-86630-665-3

Primary particles are the smallest entities that can be recognised by physical methods (e.g. optical or electron microscopy). They can be single crystals, but are for the most part composed of several crystallites with coherent lattice regions and the usual lattice defects. Their relatively simple shape may vary within a wide spectrum (isometric, cubic, aciculate, rhomboidal, irregular etc.).

Aggregates consist of groups of primary particles attached together at their surfaces. Their total surface is thus markedly smaller than the sum of the surface areas of the primary particles. Due to the high adhesion forces between the crystallites, aggregates cannot be divided into their components by comminution and they have no internal surface capable of adsorption.

The action of van der Waals or Coulombic forces causes particles and aggregates to coalesce at the corners and edges and form **agglomerates**. The surface of agglomerates differs only slightly from the sum of the surfaces of their components and is largely accessible to adsorption. Agglomerates can be broken down by comminution into aggregates and/or minor agglomerates [9].

Electron micrographs illustrate practical examples of the model just explained. Figure 1.2a shows a copper phthalocyanine of the β-form (C.I. Pigment Blue 15:3)[1] and Figure 1.2b shows a lead chromate (C.I. Pigment Yellow 34). Aggregates and agglomerates can be well discerned.

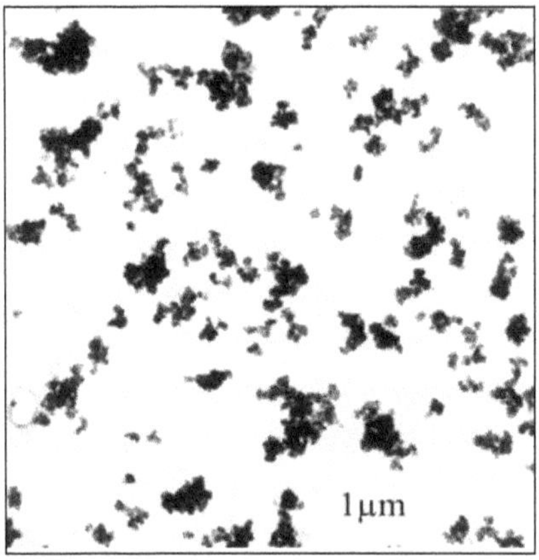

Figure 1.2a: Examples of pigment particles: Copper phthalocyanine blue, β-modification

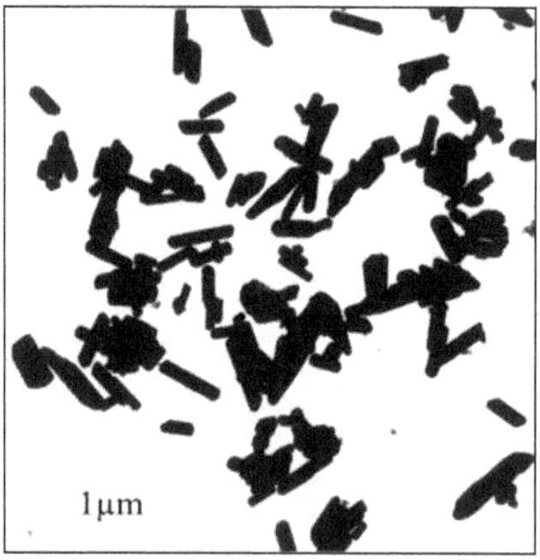

Figure 1.2b: Examples of pigment particles: Monoclinic lead chromate

Source: Works micrographe Hoechst AG

1 C.I. is the abbreviation for Colour Index, a list of pigments and dyestuffs edited by the *Society of Dyers and Colours*, Bradford, England and the *American Association of Textile Chemists and Colourists*, Lowell, Massachusetts, USA. In this index, pigments and dyestuffs are classified according to their generic names and application fields (Part 1) as well as to their chemical constitution (Part 2). See also DIN EN ISO 18451-2 for the classification of pigments for the colour index of pigments

A strict distinction must be made between agglomerates and **flocculates**. The latter are conglomerates formed of aggregates and crystals partly wetted by the binder solution and adhering more or less loosely together. They generally originate in systems with a low viscosity and can be separated again by low shear forces.

The stability of agglomerates depends upon the number of contact sites between the particles and their attractive force. Although smaller particles attract each other less, they can pack more densely so that their agglomerates may be stronger. Large agglomerates tend to break more easily than small agglomerates so that, in the course of a dispersion, the fineness of grind goes down.

1.2 Crystal structure

1.2.1 Crystal modifications: polymorphism

Apart from a few isolated cases, all pigments possess a crystal structure. Inorganic pigments consist of ionic crystals. Their lattice elements are positively and negatively charged ions, whose interlocking structure in the crystal lattice is based on the attraction between opposed electrical charges. Organic pigments are made up of molecular crystals. Their structural units are neutral molecules held together by the comparatively weak van der Waals forces.

As with crystals of other materials, pigment crystals are not perfect; i.e. they show dislocations and other flaws in both interior and surface. From the energetic point of view, faulty crystal structures are less stable than flawless crystals and this can have a negative influence on some application properties – for example, dispersibility.

Depending on conditions of crystallisation or other external factors, the molecules or ions of many organic and inorganic pigments can form several arrangements in the crystal lattice. This gives rise to different crystal phases, also known as crystal modifications. This phenomenon is called **polymorphism**. Thus, for example, quinacridones have three and phthalocyanine blue five crystalline forms. Examples of polymorphic inorganic pigments are titanium dioxide and lead chromate, each having three modifications.

The crystal modifications of a pigment have different thermodynamic stabilities. Unstable modifications can always change into the energetically more stable phase when this is favoured by external conditions. Since crystal phases may differ in their physical properties, this transmutation from one phase to another also generally results in a change in the application properties. Thus, for example, the anatase form of titanium dioxide, unlike the rutile form, is not weather resistant.

Furthermore, the rutile modification has a higher specific gravity, a higher refractive index and, thus, a better lightening power.

A further example is the case of phthalocyanine blue, whose α-form transforms into the more stable β-modification in the presence of aromatic hydrocarbons. This conversion is associated with a decrease in tinctorial strength and a change of shade towards green. Nevertheless, pigment chemists succeeded in stabilising the α-modification by the introduction of chlorine atoms into the copper phthalocyanine blue molecule. In this way, the activation energy required for the polymorphic change is increased to such an extent that the transmutation, even in the presence of aromatic hydrocarbons, is no longer possible.

Unsubstituted, linear trans-quinacridone (C.I. Pigment Violet 19) occurs, as mentioned above, in three crystal phases. Unlike the modifications of copper phthalocyanine blue, the α-, β- and γ-modifications of unsubstituted, linear trans-quinacridone differ considerably not only in their tinctorial but also in their fastness properties such as resistance to migration and fastness to light, weather and solvents. Other interesting products in this class of pigments are mixed crystals of quinacridone with substituted quinacridones or quinacridone quinone [2].

Further physical properties and characteristics that may be determined by the crystal phase are particle size distribution, particle shape, hiding power, melting point and rheological behaviour.

1.2.2 Isomorphism and anisotropy

When the crystal structures of two different substances are identical or very similar, we have a case of **isomorphism**. Isomorphic substances are capable of growing together in a common crystal lattice and thereby give rise to mixed crystals. This fact enables the manufacture of products having special properties and effects that are absent in the pure pigments. Such products are called **mixed-phase pigments**. Thus, it is possible to achieve a change of shade from yellow to red by incorporating lead molybdate into the lattice of lead chromate. Further examples are cadmium red, a mixed crystal of cadmium sulphide and cadmium selenide, and the rutile mixed-phase pigments, i.e. new coloured products that were obtained by introducing nickel and chrome oxides into the rutile lattice of titanium dioxide.

Pigments made up of anisometric crystals show anisotropic behaviour or, in other words, their mechanical, optical and electrical properties vary with direction. Forms of anisotropy are **dichroism** and **pleochroism**, as is termed the orientation dependent absorption of differing wavelengths according to the number of resulting spectra. For example, monoclinic chrome yellow crystals, which exhibit

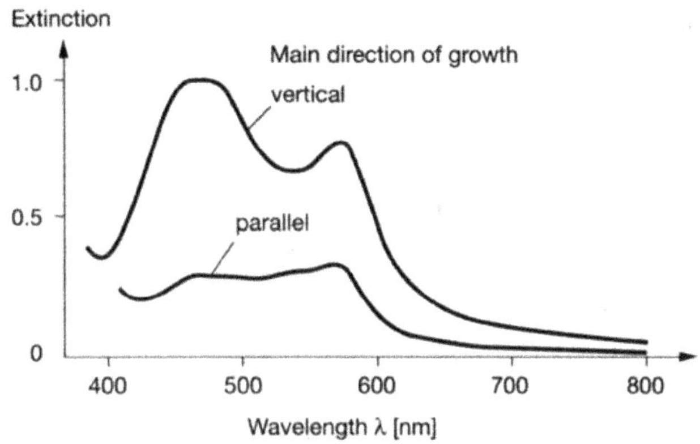

Figure 1.3: Extinction spectra parallel and vertical to the main direction of growth of a perylene pigment

Source: G. Benzing et al., Pigmente und Farbstoffe für die Lackindustrie, Expert Verlag, Grafenau, 1992

pleochroism, show varying effects depending on light incidence, the angle of observation and on their alignment in the medium. The acicular copper phthalocyanine blue is another example of the dependence of absorption spectra upon the radiation path. They differ in the frequency of the absorption maxima, resulting in a change of shade towards reddish blue in reductions. For example, Figure 1.3 shows the extinction spectra of a perylene pigment (C.I. Pigment Red 179) measured at right angles and parallel to the direction of crystal growth.

1.2.3 X-ray crystallography

The structural elements of crystals – atoms, ions or molecules – form spatial lattices with constants, i.e. the distance between the planes, of the same magnitude as the wavelength of X-rays (0.01 to 10 nm). On passing an X-ray beam through such a lattice, the regularly arrayed units form scattering centres and interference effects occur. The X-rays are diffracted by the crystal lattices in the same manner as visible light is diffracted by a plane diffraction grating, and the diffraction patterns thus obtained are characteristic of the diffracting crystal lattice (*Max von Laue*, 1912). This phenomenon underlies all methods for the X-ray analysis of crystals (*Laue, Bragg, Debye-Scherrer*). The **Debye-Scherrer method** is particularly important for pigment characterisation.

The interference or diffraction pattern results as follows. The elementary waves originating in the structural units of the crystal lattice may, depending on phase displacement, interfere constructively or destructively, giving rise to either reinforcement, attenuation or extinction. The so-called **Bragg's condition**:

$$2\,d\,\sin\vartheta \;=\; n\,\lambda \qquad (n \;=\; 1, 2, 3\,...)$$

states that reinforcement of two waves originating from the same beam can only take place when their path-length difference ABC (Figure 1.4) is equal to the wavelength or to an integral multiple of it. This is the case when the incident beam of X-rays strikes the crystal plane at an angle ϑ. This angle is termed **Bragg's glancing angle**, d is the interplanar distance and n the arbitrary integer which specifies the **order of reflection**.

The powder method of *Debye* and *Scherrer* is the usual technique for the structural determination of pigment crystals. A scheme of the experimental arrangement is illustrated in Figure 1.5. Instead of a single crystal, a sample is used which is made up of finely powdered crystals compressed to form a small rod. Since the orientation of the crystals in the sample is totally random, there will always be some crystals with their lattice planes in a position satisfying the *Bragg* condition when struck by an incident beam of X-rays. Such an arrangement makes it unnecessary to rotate the sample, as is the case in Bragg's single crystal method. The "reflected" radiation spreads out conically around the incoming beam and results, in the case of photographic recording, in a pattern of circles with a central spot due to the undiffracted beam on a strip of film wrapped concentrically around the sample. The resultant diffractogram is practically an identity card for the examined crystal modification, i.e. of the corresponding pigment.

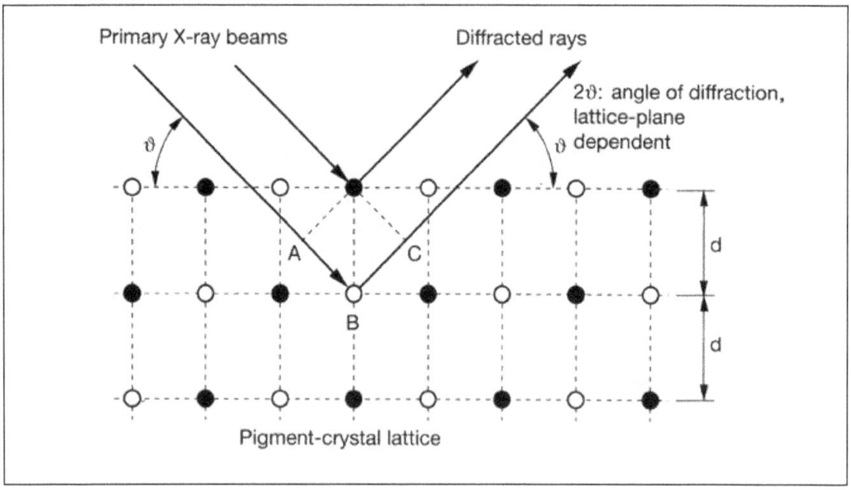

Figure 1.4: Reflection (and diffraction) of monochromatic X-rays at crystal planes according to the Bragg conditions

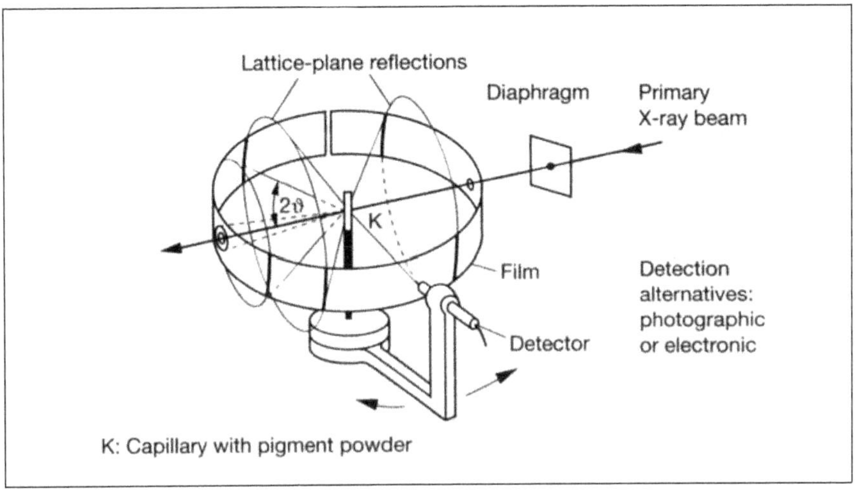

Figure 1.5: Experimental arrangement for the powder method of Debye and Scherrer
Source: P. Atkins Physikalische Chemie, VCH Verlagsgesellschaft, Weinheim, 1988

In modern diffractometers the diffraction pattern of the powdered sample is recorded electronically, by means of a suitable X-ray detector.

In comparison with the single crystal methods of *von Laue* and *Bragg*, the powder method shows two advantages: Firstly, there is no need for large single crystals as samples, and secondly, the glancing angles of all lattice planes can be recorded in one operation without rotating the specimen. Further details are given in [3] and [4].

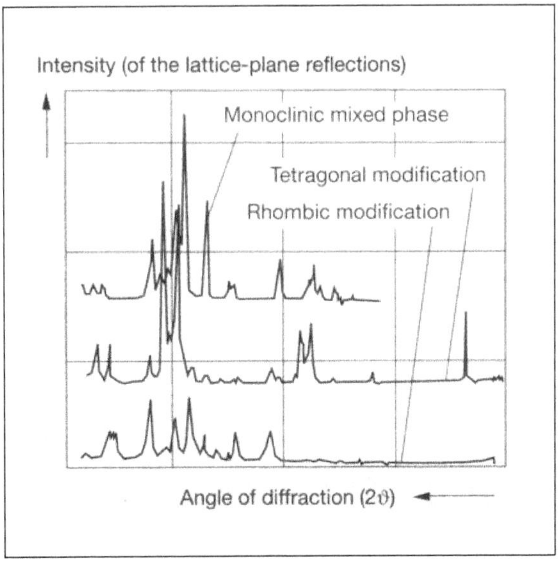

Figure 1.6: X-ray powder-diffraction patterns for the three crystal modifications of molybdate red
Source: W. Herbst and K. Merkle. Deutsche Farben-Zeitschrift, N°8, 1970, p. 365

Figure 1.6 shows by way of example the *Debye-Scherrer* diffraction diagrams obtained with three samples of mono-clinic, tetragonal and rhombic modifications of molybdate red. Only the first two modifications have attained commercial importance. Both the monoclinic and the rhombic modifications are sensitive to grinding. Thus, intensive dispersion may cause primarily a decrease in hiding power.

X-ray analysis also allows the detection of a possible isomorphism in pigments that are chemically different. Furthermore, this method is suitable for the examination and control of mixed-phase pigments.

1.3 Particle size measurement

Being technical products, pigments do not form in a uniform particle size, but consist of different-sized particles. They are said to be **polydisperse**. The particle size of organic pigments varies between 0.01 and 0.1 µm, that of inorganic pigments being about twice as high. Titanium dioxide may even contain particles of up to 1 µm. Further exceptions are Prussian blue, with a particle size range corresponding rather to that of the organic pigments, high-quality carbon blacks and transparent iron oxides, which have a particle size easily a tenth power below the range of organic pigments.

Data concerning particle size (average particle diameter, equivalent diameter, spread of distribution, particle size distribution, specific surface area) are con-

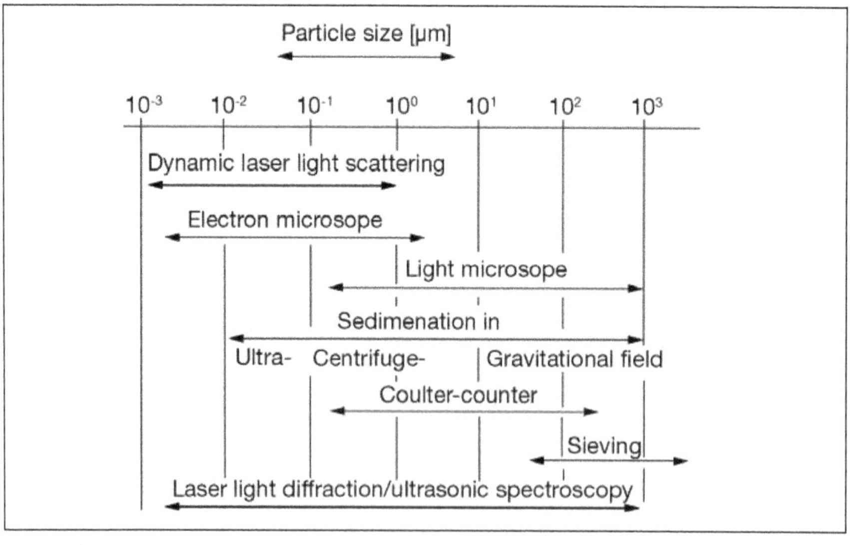

Figure 1.7: Comparison of the ranges of particle size determinable by different methods

sidered to be, together with the particle shape and optical constants, the most important data for pigments. Properties such as light absorption and scattering, shade and dispersibility as well as the rheology of the pigmented system depend upon particle size.

1.3.1 Particle size distribution

Several methods have been developed for the assessment of particle size distribution. In all cases the preparation of the sample is very important, as this has a decisive influence on the result. In particular, the dispersing medium and the energy input during dispersion play a cardinal role. On the other hand, care must be taken that the level of dispersion does not change in the course of measurement. The dispersion medium usually consists of water or organic solvents. In the case of pigments that are not readily wetted, help is provided by the addition of surfactants.

The methods most used in practice are:

Sieve analysis
Evaluation of transmission and scanning electron micrographs
Sedimentation analysis with disc or ultra-centrifuge
Optical methods
Ultrasonic attenuation spectroscopy
Electroacoustic methods
Coulter counter

A classification of these and other methods according to their relevant ranges is shown in Figure 1.7.

1.3.1.1 Sieve analysis

This method consists of the determination of particle size by means of sieves with different mesh sizes. As its measuring range only records the coarse fraction of the particle size spectrum, this process is only suitable for inorganic pigments to be used in emulsion paints. In many cases it is also of interest to determine the fractional residue of over-size pigment particles considerably exceeding the average particle size. Sieve analysis is not appropriate for reliable information on particle size distribution, all the more so for organic pigments, and is only considered here for the sake of completeness.

The procedure is as follows: The pigment powder is suspended in water, and this suspension is poured through a sieve of suitable mesh width. The sieve is flushed repeatedly with tap water until the wash water remains clear. The sieve residue is dried, weighed and given as a percentage of the original sample (DIN 53 EN

ISO 787-7). A mechanical sieve-residue determination by means of the **Mocker apparatus** (DIN EN ISO 787-18) also exists.

1.3.1.2 Determination by electron microscopy

On account of its insufficient resolving power, the light microscope cannot be used for the examination of pigment aggregates, which are as a rule smaller than the wavelength of visible light. For this purpose, resort must be made to the electron microscope, which has a lower limit of registration more than a tenth power below that of the light microscope. The increase in resolving power was possible thanks to *L. de Broglie's* theory that electron beams, when considered to be corpuscular radiation, must possess wavelike properties (1924). Hence, the interaction between electron beams and matter results in the same diffraction effects as in the interaction between light and matter. Moreover, advantage was taken of the fact that electron beams, like light radiation, spread in straight lines in a vacuum and can be focused by magnetic and electrical "lenses" just as light beams are focused by glass or quartz lenses.

Electron microscopic examination enables both a qualitative and a quantitative characterisation of pigment particles. In the first case, transmission electron micrographs make possible the determination of particle size and morphology. The quantitative evaluation, usually requiring higher expenditure, aims at the assessment of particle-size distribution. In principle, differentiation between primary particles, aggregates and agglomerates due to their distinct transparency when viewed by transmitted electron beams is feasible. However, when evaluating clusters or anisometric particles of organic pigments with automatic counting devices, an absolutely reliable discernment of individual structural units concealed in agglomerates is not always possible, thereby impairing the accuracy of the measurement. In such cases, the scanning electron microscope has generally proven to be better.

A further difficulty arises with coated pigments because, depending on their concentration, the measurement of particle size may be rendered somewhat problematic by the products coating the pigment surface.

In spite of these limitations, the importance of information afforded by electron microscopic methods is beyond debate. Nowadays they have become an essential help for characterising pigment particles. Their particular advantage in comparison with sedimentation methods is the possibility of determining particle size distribution in the medium of application, e.g. in the coating layer.

1.3.1.3 Analysis by sedimentation techniques

Sedimentation methods are based on *Stokes'* law, which states the relationship between the free falling velocity of a spherical particle in a Newtonian fluid (see

Section 3.4.1) and the particle diameter. If this is d, the particle density ρ_1, the density of the fluid ρ_0 and the gravity constant g, then the downward force F_1 is given by the following equation:

$$F_1 = \frac{4}{3}\pi\left(\frac{d}{2}\right)^3 (\rho_1 - \rho_0)\, g$$

As the particle settles through the viscous liquid, friction exerts an opposing force F_2 on the particle:

$$F_2 = 3\,\pi\, d\, \eta\, v$$

where η = viscosity coefficient and

v = sedimentation velocity

The settling velocity increases till F_2 becomes equal to the downward force. F_1 and F_2 are then balanced and the particle falls at constant velocity. By equating the two opposing forces, we obtain the expression:

$$3\,\pi\, d\, \eta\, v = \frac{4}{3}\pi\left(\frac{d}{2}\right)^3 (\rho_1 - \rho_0)\, g$$

Solving this equation for d yields Equation 1.1:

Equation 1.1:

$$d = \sqrt{\frac{18\,\eta\, v}{(\rho_1 - \rho_0)\, g}}$$

Because of the diminutive size of pigment particles (< 1 µm), settling caused by gravity takes a considerable amount of time. An acceleration of the sedimentation process by means of centrifugal forces is feasible. Thus the ultracentrifuge allows the settling of particles even in the colloidal range (0.01 to 0.05 µm).

A prerequisite for sedimentation analysis is that each particle must settle freely and independently of its neighbours. This requires concentrations in the range of 0.1 %, and thus the possibility of shock phenomena must always be kept in mind when dilution of the pigment suspension is carried out.

Whereas the sedimentation process was formerly determined gravimetrically, this is nowadays performed by means of optical methods. In this so-called **photosedimentometer technique**, a beam of light is passed through the settling suspension and the changes in transmission are continuously measured by a photocell. The transmission values allow calculation of the extinction, which can be carried out on the settling suspension in only one operation. The particle size distribution can be ascertained from the extinction values with aid of Equation 1.1, which was derived from *Stokes'* law, and the relationship between the relative extinction area, the optical constants of the pigment and the refractive index according to the theory of *Mie*. Both white and monochromatic (e.g. laser light) radiation are suitable for the transmission measurements [5].

Critics of the sedimentation method have pointed out that Stokes' law is strictly speaking only valid for spherical particles. The sedimentation behaviour of non-spherical particles has not been definitely determined as yet. Nevertheless, in the opinion of experts, the approximation to spherical shape is justified by the Brownian motion of the particles during settling.

Details on the principles and performance of ultra-sedimentation can be found in the relevant literature [6–8].

1.3.1.4 Optical methods

The most frequently used methods for determining mean particle sizes and particle size distributions rely on the measurement of scattered or diffracted light from the particles. Mainly, two different methods are used. One is **dynamic light scattering** (DLS) and the other is **laser light diffraction** (LLD).

When light hits dielectric objects that are in the size range of the wavelength of the light, it is scattered. The reason for this is the interaction of the electric field of the light rays with the electrons in the objects, causing them to oscillate. The electrons then send out the energy once again, similar to radio antennas. An oscillating electron cannot send radiation out in the direction of its oscillation, but only perpendicular to it. Therefore the light is partly polarized, depending upon the angle from which it is observed [180]. The direction and the intensity of the scattered light depend upon the refractive index and the absorption coefficient for each wavelength as well as on the size of the particles. Both in LLD and DLS, lasers in the visible part of the spectrum are put to use as light sources.

Dynamic light scattering (DLS)

Dynamic light scattering is also called **photon correlation spectroscopy** (PCS) or **quasielastic light scattering** (QELS). This method is used to measure particles in the size range between 1 nm and 1 µm. The motion of suspended particles of this size range in liquids is determined by the Brownian motion within the liquid. The **Stokes-Einstein** equation relates the so-called **translational diffusion coefficient** with both the viscosity of the liquid and the particle size. By measuring the change in scattered light intensity over very short time periods, the diffusion coefficient is determined. Apart from a mean particle size (Z-average) which is related to the intensity of the scattered light, a **polydispersity index** is generated that relates to the spread of the particle size distribution. The Z-average can be transformed into a volume average (median) or a number average (mean). Since it is only necessary to determine the fluctuation in the scattered light intensity, the detector is placed only at one angle, normally at 90° to the incident light.

The particle concentration during the measurement is typically very low. Otherwise, the particles would obstruct their diffusion paths by bumping into each other. Due to the poor light scattering ability of minute particles and their low concentration, the detected signal is very low also and any strong signal that might come either from a large particle or, for example, from a speck of dust will render the measurement meaningless. DLS is not only used to determine the size of nano-scaled particles, but also to establish the sizes of emulsions, polymers and micelles.

It should be kept in mind that any adsorbed material on the surface of the particles will change their diffusivity. Therefore, the measurements might overestimate the particle size because the **hydrodynamic** volume may be larger than the geometrical one.

Laser light diffraction (LLD)

Laser light diffraction, sometimes also referred to as **low angle laser light scattering** (LALLS) is used for measuring particles in the size range between 30 nm and 1 mm. When coherent light is directed onto a particle, it is diffracted. Behind the particle the diffracted light can interfere so that a series of maxima and minima will appear, depending upon if the interference was constructive or destructive. Since smaller particles diffract the light to a larger extent (the angle of aperture is larger), the maxima and minima are farther apart from each other than for larger particles and also less pronounced.

In the case of LLD, the light is either passed through a suspension of particles or through a cloud of particle dust. The pattern that is generated is analysed in order to obtain information on the particle size distribution. The basic requirement for instruments and for the method as a whole is described in the standard ISO 13 320.

When the method was first utilized in the late 1970s, computers were unable to evaluate the diffraction patterns using, for example, the *Mie* theory of light scattering because random access memories were insufficient. Therefore, Fraunhofer diffraction algorithms were employed. This was fine for rather course particles, but not for the size range that pigments and fillers normally cover. Therefore, the particle sizes thus determined were, as a rule, too large. In the meantime, computation ability is no longer a limitation and the Mie theory, which covers the size range from the nano-scale onwards, is used for evaluations. In fact, the new ISO 13 320 states that the Fraunhofer evaluation method is only feasible for particles larger than 40 times the wavelength of the laser light used. So, when using for example a He-Ne-Laser with a wavelength of 633 nm, the lowest recommended particle size would be 25 µm.

1.3.1.5 Ultrasonic attenuation spectroscopy (ultrasound spectroscopy)

In ultrasonic attenuation spectroscopy, an ultrasonic wave with a power in the order of several tens of milliwatts is passed through a suspension of particles and the attenuation α is measured as a function of the frequency and the distance between the source and a detector. The attenuation is given by

$$\alpha = \frac{10}{f \cdot L} \lg\left(\frac{I_0}{I}\right)$$

It has the physical dimension dB/cm MHz. In the above equation, f is the frequency and L the distance between the transducer and the detector. I_0 and I are the intensities of the generated and detected ultrasonic waves.

Ultrasonic waves passing through suspensions of particles in a liquid are attenuated by a number of processes. Firstly, there is an intrinsic absorption α_{int} of the energy of the waves by the liquid phase, since the compression of liquids cause dissipation, meaning that the liquid gains heat. If the particles are compressible, then this will also cause a loss of energy of the ultrasonic waves. This is called thermal attenuation α_{th} and is a major influence when measuring emulsions, for example. Particularly when the particles are rigid, the sound waves can be diffused by scattering, leading to a loss α_{sca}. In ultrasonic spectroscopy this component becomes negligible if the particles are smaller than, say, 5 µm. Should there for some reason be a structural interaction between the particles due to, for example, a high concentration, then a component α_{st} is to be considered. Lastly, there is a loss of energy due to shear and friction at the interface between the particles and the liquid phase, since the particles are accelerated and caused to move back and forth within the liquid by the ultrasonic waves. This component of attenuation is called **visco-inertial**, α_{vis}. The total attenuation measured is the sum of the individual parts.

$$\alpha_T = \alpha_{\mathrm{int}} + \alpha_{sca} + \alpha_{h} + \alpha_{s} + \alpha_{vis}$$

Using mathematical models and background measurements, it is possible to determine or estimate the individual components and determine a particle size distribution from the data [181]. The advantage of using ultrasonic attenuation spectroscopy to determine the particle size distribution is seen in the fact that concentrated suspensions can be measured. Compared to laser light diffractions techniques, ultrasound spectroscopy gives more detailed information especially on the particle fractions with less than 0.1 μm diameter.

1.3.1.6 Electroacoustic methods

Particle size measurements can be carried out in combination with measuring Zeta potentials using electro-kinetic sonic amplitude (ESA) or the ultrasonic vibration potential (UVP) techniques. These methods are described in Section 4.2.3.4, "Measuring zeta potentials".

1.3.1.7 The Coulter counter

A strongly diluted dispersion of the pigment in an electrolyte has to be prepared for the assessment of particle size distribution with the Coulter counter. This suspension is allowed to flow through a narrow orifice situated between two electrodes. During passage through the aperture, each particle displaces its own volume of the electrolyte. Since the conductivity of the particles differs from that of the electrolyte, the volume displacements give rise to deviations in the conductivity that are proportional to the particle volume. These changes in conductivity are registered by means of an electronic counter.

This method is suitable for particles of 0.2 μm and less. Opinions differ as to its accuracy. It has been stated that extremely small particles remain undetected, particularly when the particle size spectrum is very narrow. This is attributed to a "shading" effect of the bigger particles. On the other hand, the possibility of pigment re-agglomeration during dilution must be taken into consideration [9, 10].

1.3.1.8 Methods of presenting results

The results of empirically determined particle size distributions can be presented in various ways, such as:

Tables
Histograms
Curves
Non-linear representations such as power-function and logarithmic grids for sum distributions

In practice, graphical methods are preferred due to their better clarity (DIN ISO 9276-1). When the result consists of ten or less measurements, then a histogram is the proper procedure. In this case the relative size frequency is plotted in rectangular bar form versus a specific particle size parameter, e.g. equivalent diameter (Figure 1.8a). In case of more than ten measurements being available,

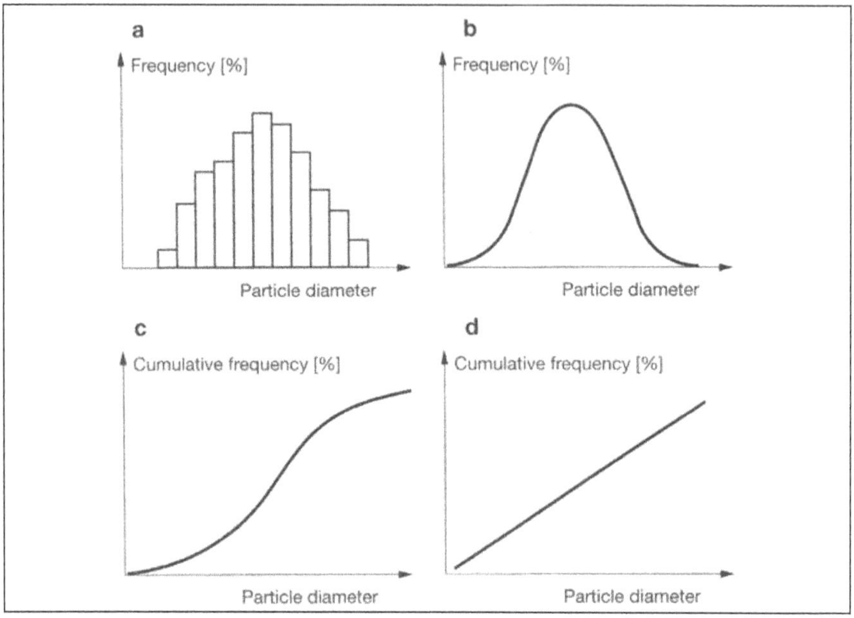

Figure 1.8: Representation of particle size distribution: (a) histogram, (b) distribution curve, (c) cumulative distribution, (d) cumulative distribution on probability grid

the representation as a curve could be more convenient (Figure 1.8b). The **sum distribution**, which shows the cumulative size frequencies up to a given size as a function of the corresponding equivalent diameters (Figure 1.8c) is also often used. Plotting such sum distributions on non-linear grids (power, logarithmic or probability grid) enables easy inter- and extrapolations (Figure 1.8d). There are special standards for some of these representations, e.g. power-function grid (DIN 66 143) and log-normal grid (DIN 66 144).

Frequency as well as cumulative distributions can be presented as relative number, volume, mass or surface area distributions plotted against the equivalent diameter. Ultimately, the manner of presenting the results will depend on the one hand on the method of determination, and on the other hand on the purpose for which the determination was carried out. Thus, for example, results from the evaluation of electron micrographs are best interpreted if presented as numerical distribution. When coloristic points of view are the most important criteria, mass or volume distributions should be preferred, as the extinction of the pigment particles, which plays the decisive role here, is proportional to their mass. When studying particle size distribution with a view to dispersibility, presentation of the results as a surface-area distribution is more informative because pigment dispersibility is connected with its surface energy.

An example for the result of particle size distribution analysis is the bell-shaped normal or *Gaussian* distribution (Figure 1.8b). However, the particle size distributions of pigments, especially organic pigments, are very often unsymmetrical. In such cases, it is better to resort to log-normal distribution paper (DIN 66 144). When the particle size distribution in question corresponds to a log-normal distribution, its cumulative distribution on this kind of paper results in a straight line.

1.3.2 Specific surface area

The specific surface area is defined as the surface area per weight unit of the pigment. It is usually given as square metres per gram. The mean particle diameter (in the case of spherical particles) or the mean equivalent diameter (when the particle shape diverges from the spherical) can be derived from this value. In addition, the specific surface area is, as a rule, an important concept for characterising pigments, as their application behaviour is strongly related to their surface area and its properties.

Due to the fact that pigment particles have without exception an extremely small size, the specific surface area can only be measured indirectly. The standard method for this determination is the one developed by *S. Brunauer, P. H. Emmett* and *E. Teller* (BET method).

The absorption method is based on the following principles: Molecular forces at the surface of a solid (called *adsorbent)* are not entirely saturated as is the case with forces in the bulk. The surface forces basically lack the bonding partner. For this reason, they are in a position to attract molecules of a gas (called *adsorptive*) and to temporarily adsorb them. Depending on the kind of forces giving rise to the adsorption process, a difference is made between **physisorption** and **chemisorption**. Physisorption is caused by *van der Waals* attraction forces, i.e. dipolar and London or dispersion forces, and also ionic forces and hydrogen bonds. Chemisorption results when the bond is due to chemical interaction between the solid and gas molecules. Chemisorption is more stable than physisorption as an actual chemical reaction takes place.

In contrast with physisorption, which depends on pressure and is always reversible, the chemisorption process is independent of pressure and partly irreversible. A sharp distinction between both kinds of adsorption is, however, not always possible.

1.3.2.1 The Freundlich adsorption isotherm

In a solid/gas system the amount of adsorbed gas (called *adsorbate*) depends on pressure and temperature. For the particular case of monolayer coverage at constant temperature, *H. Freundlich* was able to describe this dependence by the empirical relation:

Equation 1.2:

$$a = k \cdot p^{\frac{1}{n}}$$

where

a = amount of adsorbate per weight unit of the adsorbent (in g),

p = (partial) pressure (in bar)

k, n = specific constants which depend upon temperature, type of gas and adsorbent

This relation, which is valid only for a limited pressure range, is based on the supposition that adsorption energy decreases exponentially with growing coverage of the solid surface. The value of constant n varies between 1.5 and 5. The presentation of the amount of adsorbate as a function of pressure at constant temperature is called an **adsorption isotherm**. The *Freundlich* adsorption isotherm should give a straight line when log a is plotted against log p.

1.3.2.2 Langmuir's adsorption theory

According to *I. Langmuir's* molecular kinetic conception of the adsorption process, gas molecules condense onto the solid surface and, at the same time, a number of adsorbed molecules become free. At a given pressure and constant temperature this process reaches a dynamic equilibrium, i.e. the number of molecules newly adsorbed equals the number of molecules leaving the adsorbent. At this stage, according to Langmuir's assumptions, the adsorption does not proceed beyond single-layer coverage. This is due to the fact that, after formation of the monolayer, the adsorption forces are saturated so that the number of adsorbed molecules cannot increase further.

Denoting (in conformity with standardisation) 1 as the total solid surface and θ that part of the adsorbent which is covered with the adsorbate, then $(1 - \theta)$ represents the uncovered part of the adsorbent surface. The rate of adsorption is proportional to both pressure and the uncovered surface. On the other hand, the rate of desorption is proportional to the covered part. Thus:

$$\text{rate of adsorption} = k_a \; p \, (1 - \theta),$$

$$\text{rate of desorption} = k_d \; \theta$$

At equilibrium both rates will be equal, so that:

$$k_d \; \theta = k_a \; p \, (1 - \theta)$$

Solving for θ gives:

$$\theta = \frac{k_a p}{k_d + k_a p}$$

The amount of adsorbate per surface unit is proportional to the covered surface, i.e. $a = k\,\theta$, with k as new constant of proportionality. With a constant k_2 equal to k_a/k_d, and another constant k_1 equal to k_2 k, we obtain Equation 1.3.

Equation 1.3:

$$a = \frac{k_1 p}{1 + k_2 p}$$

The relationship expressed by Equation 1.3 is termed the **Langmuir adsorption isotherm**. It describes a hyperbola beginning at the origin (Figure 1.9). Initially, as long as $k_2 \ll 1$, the amount of adsorbate is proportional to pressure. At higher pressure, $k_2 \gg 1$, the amount of adsorbate approaches a maximum value that is attained when the surface is saturated and remains constant. Starting from this point, the hyperbola runs parallel to the abscissa.

The Langmuir adsorption isotherm can also be expressed in the form:

$$\frac{p}{a} = \frac{1}{k_1} + \frac{k_2}{k_1} p$$

From this expression it is easy to deduct that plotting the quotient *p*/a as a function of pressure *p* results in a straight line (**Figure** 1.9).

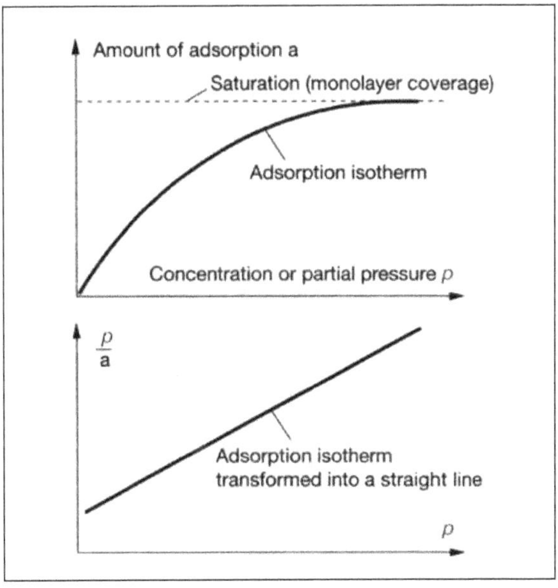

Figure 1.9: Adsorption isotherms according to Langmuir (above), transformation into a linear relationship by changing the co-ordinates (i. e. p/a versus p) (below)

1.3.2.3 The adsorption isotherm according to Brunauer, Emmett and Teller

Closer examination revealed that the *Langmuir* treatment of a single molecular layer explained only some cases of adsorption and could not be generalised. In most cases, adsorption measurements resulted in S-shaped isotherms. *Brunauer, Emmett* and *Teller* contrived an improved theory [11]. Unlike the *Langmuir* approach, they assumed that after formation of a monolayer of the adsorbate and the consequent saturation of the adsorption forces, the van der Waals cohesion forces were by themselves sufficient to stack subsequent layers. Thus a capillary condensation in the adsorbent pores would take place. This process would be particularly favoured by low temperatures and by pressures near the vapour pressure.

Figure 1.10 above shows the isotherm of such a multilayer adsorption with its characteristic S-shape. The turning-point B corresponds to completion of the monolayer; the following rising part of the curve indicates the capillary condensation. Thus a first approximation for the so-called **monolayer capacity**, i.e. the necessary number of molecules for monolayer formation, can be obtained directly from the isotherm plot.

For the gas volume V_{ads} adsorbed in the course of a multilayer adsorption, the following equation could be derived:

$$V_{ads} = \frac{V_m \, c p}{(p_s - p)\left[1 + (c-1)\dfrac{p}{p_s}\right]}$$

where

V_m = volume of the monolayer at STP (standard temperature and pressure)
p = partial pressure of the adsorbed gas at equilibrium
p_s = saturation vapour pressure of the adsorbed gas at experimental temperature

c = a constant > 1, which expresses the rate of interaction between adsorbate molecules and adsorbent surface. Its value corresponds to exp $\Delta H/RT$; ΔH being the difference between the adsorption and the vaporisation enthalpies of the monolayer, R the molar gas constant and T the thermodynamic temperature.

This equation may be rearranged into the form:

Equation 1.4:

$$\frac{p}{V_{ads}\,(p_s - p)} = \frac{1}{V_m \, c} + \frac{c-1}{V_m \, c}\,\frac{p}{p_s}$$

which shows that plotting $p/V_{ads}(p_s - p)$ against p/p_s gives a straight line. From its slope and its intersection with the ordinate axis, c and V_m can be calculated (**Figure** 1.10 below).

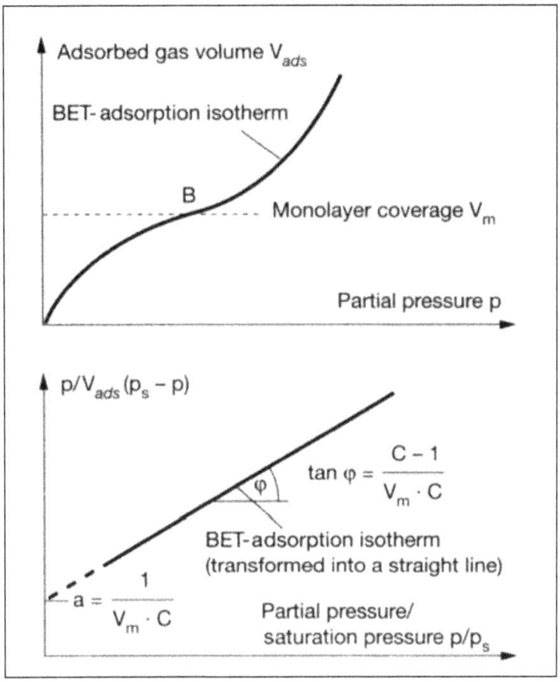

Figure 1.10: BET-isotherm for multi-layer adsorption (above); transformed into a linear relationship (below)

Equation 1.4 is the most common form of the BET equation. Adsorption isotherms that can be evaluated in accordance with it give in most cases a straight line for the p/p_s range between 0.05 and 0.35. For $p/p_s < 0.05$ the obtained V_{ads}-values are too low and for $p/p_s > 0.35$ they are too high. Despite this limitation, innumerable measurements since the development of the BET theory have shown good conformity with values obtained by other methods. Deviations lie in the magnitude of a few percent.

Since V_m and c are characteristic constants for each adsorbate/adsorbent system, the corresponding fractions in Equation 1.4 namely:

$$\text{the slope } \frac{c-1}{V_m \cdot c} \quad \text{and the intercept } \frac{1}{V_m \cdot c}$$

are also constants. With b as the slope (tan φ) and a as the intercept (**Figure** 1.10 below), we can derive for V_m and c the expressions:

$$V_m = \frac{1}{a+b} \quad \text{and} \quad c = 1 + \frac{b}{a}$$

The monolayer capacity V_m and the area per molecule of the adsorbed gas (0.162 nm^2 for nitrogen) allow us to calculate the specific surface area. In the case of mere monolayer coverage, the BET equation reduces to the *Langmuir* equation.

1.3.2.4 Determination of the average particle size

Either the average particle diameter d or the average equivalent diameter d_E can be determined with help of a simple relationship between specific surface area S and d. The volume of a spherical pigment particle with density ρ and diameter d is 1/6 π d^3. Thus, for the number of particles N in 1 g:

$$N = \frac{6}{(\pi d^3 \rho)}$$

The surface area of the particle being π d^2, it follows that for the specific surface area S:

$$S = \frac{6}{(\pi d^3 \rho)} \pi d^2 = \frac{6}{d \rho}$$

and hence for the above-mentioned relationship:

Equation 1.5:

$$d = \frac{6}{S \rho}$$

This relationship for calculation of the true average particle diameter applies only to isometric (spherical or cube-shaped) particles. For non-spherical particles, the **average equivalent diameter** d_E is determined, i.e. the diameter of spherical particles with the same volume. For this the numerator 6 in Equation 1.5 must be replaced by a so-called **sphericity value**, which takes into account the deviation of the particle shape from the spherical. This sphericity value is 4 instead of 6 for acicular and 2 for flaky particles, as occurs in the case of effect pigments. Intermediate stages are also possible.

Specific surface area values and the corresponding average equivalent diameters for some pigments are given below (Table 1.1).

Table 1.1: Specific surface area in m²/g and equivalent diameter in µm of several pigments

Pigment Type	C.I. denomination	C.I. number	Specific surface area [m²/g]	Equivalent diameter [µm]
Hansa yellow	Pigment Yellow 10	12710	16	0.250
Thioindigo	Pigment Red 88	73312	30	0.100
Cu-phthalocyanine	Pigment Blue 15	74160	62	0.060
Carbon black	Pigment Black 7	77266	180	0.020

In order to avoid a possible chemisorption during the adsorption process, inert gases such as nitrogen or argon are used in practice. A prerequisite for reliable readings is that the gas must have access to the whole internal surface of the pigment, which is not always the case. Very finely divided or strongly agglomerated organic pigments are particularly critical, as the internal surface of agglomerates may be unattainable for the gas molecules. This results in lower readings being obtained. Coated pigments present a further difficulty as the added substances, e.g. resins, may falsify the data partly due to their own porosity and partly owing to the presence of separated particles. It is thus always advisable to view information on specific surface area in connection with other physical data of the pigment.

It has already been mentioned that, when determining specific surface areas, only the external surface of agglomerates and aggregates as well as the internal surface of the agglomerates are covered, because only these surfaces are accessible to adsorption. On the other hand, aggregates are as a rule not significantly comminuted during conventional dispersing processes. Thus, it is generally accepted that pigment dispersion predominantly consists in dividing agglomerates into aggregates. This means that the pigment surface area covered by the BET method is the same surface that occurs afterwards in the disperse system – provided that thorough dispersing has taken place – when the particles have been wetted by binder and solvent. Since the surface area of the adsorbed and immobilised binder layer is directly proportional to the pigment surface accessible to adsorption, it follows that the specific surface area gives a reference point for the flow properties of the pigmented system (see Section 3.5.1). In addition, a study by *Apel* has shown that a strong connection also exists between specific surface area and optical properties such as tinting strength and hiding power [12].

Standards DIN 66126-1, 66126-2 and 66 131: deal with the determination of specific surface area. Detailed descriptions and theoretical principles can be found in the relevant literature [9, 13].

Example 1

The data below was obtained from the adsorption of nitrogen

Table 1.2: Co-ordinate values for the linear transformation of the adsorption isotherm in Example 1 using the BET-Method
p = pressure at equilibrium, p_s = vapour pressure at saturation, V_{ads} = adsorbed gas volume

$\frac{p}{p_s}$	$\frac{p}{V_{ads}(p_s - p)}$ [1/ml]
0.051	0.0048
0.076	0.0071
0.093	0.0087
0.123	0.0114
0.159	0.0148
0.209	0.0193
0.242	0.0223
0.320	0.0294

on 0.55 g of dioxazine violet (C.I. Pigment Violet 23) at 77 K (p is the pressure and V_{ads} the adsorbed volume at STP):

p in [kPa]:	5.2	7.7	9.4	12.5	16.1	21.2	24.5	32.4
V_{ads} in [ml]:	11.3	11.6	11.8	12.4	12.8	13.7	14.3	16.0

The saturated vapour pressure p_s of nitrogen is 101.3 kPa. The molecular area of nitrogen is $16.2 \cdot 10^{-20}$ m². Calculate the specific surface area of dioxazine violet.

Solution

We start by employing the data to determine the co-ordinate values for the linear relationship of the isotherm according to the BET method (Equation 1.4). Table 1.2 is drawn up from the result.

The slope of the straight line, b, and the intercept at the ordinate axis, a, can either be read from the BET plot or calculated by linear regression (method of Least Squares). The linear regression, accomplished with help of a desktop computer, gave the following values:

$$a = 0.000174; \quad b = 0.091444$$

Therefore V_m is:

$$V_m = \frac{1}{0.091444 + 0.000174} = 10.91 \text{ ml}$$

and for 1 g pigment:

$$V_m = 10.91 : 0.55 = 19.84 \text{ ml}$$

Using the Avogadro constant $6.02 \cdot 10^{23}$ (number of molecules in 1 mole of any substance) and with the help of the fact that the molar volume of an ideal gas under STP conditions (i.e. at 0 °C and 101,325 Pa pressure) is 22,400 ml, we obtain for the number of molecules N in 19.84 ml of nitrogen under STP conditions:

$$N = \frac{19.84 \cdot 6.02 \cdot 10^{23}}{22400}$$

$$N = 0.005332 \cdot 10^{23}$$

Hence, it follows for the specific surface area per g of adsorbent:

$$S = 0.005332 \cdot 10^{23} \cdot 16.2 \cdot 10^{-20} \text{ m}^2\text{/g}$$

$$S = 86 \text{ m}^2\text{/g}$$

1.4 Surface character

Besides particle shape and specific surface area, there are other physical properties of the pigment, particularly concerning its surface, that exert a decisive influence on the interaction between the pigment and its environment. These properties include polarity, hydrophilicity, surface tension, number and type of active groups on the particle surface and polarisability of the pigment molecule. Such properties are summarised under the concept **surface character.**

As a rule, the surface character is determined by examining the interaction of the pigment with media chosen for this purpose (binders, solvents, surfactants). The molecular structure itself provides first indications as to the interaction to be expected. Thus, for example, the high hydrophilicity of perylene red is caused by the presence of polar acid-imide groups on the periphery of the molecule. A further, good example is titanium dioxide, where the oxygen ions bestow a marked polar, i.e. hydrophilic, surface on the particles. On the other hand, the strong hydrophobic character of the phthalocyanine blue molecule can be attributed to the non-polar benzene rings at its surface.

1.4.1 Polarity and hydrophilicity

The degree of polarity of a pigment surface depends on its chemical constitution. When the pigment molecule has a dipole moment, this will result in a pronounced polar character and the pigment is considered as hydrophilic. In other cases, i.e. for molecules without a dipole moment, a non-polar surface is present and the pigment is termed hydrophobic or organophilic. Many inorganic pigments, especially those of the salt or metallic oxide types, are **a priori** polar on account of their ionic structure or the high electron affinity of the oxygen atom. Organic pigments have a polar surface if polar functional groups like hydroxyl, carboxylic or carbonyl groups are present on the periphery of their molecules, or if they contain atoms with high electronegativity. It is also possible to influence the surface polarity of a pigment by after-treatment (see Section 1.4.4).

A reference for the polarity of a pigment surface is supplied by the **hydrophilicity** as defined by *Zettlemoyer*[14], and for this purpose the BET water vapour isotherm is determined. Assuming that the area per water molecule is 0.105 nm^2, the proportion of the surface area covered with water can be calculated. The hydrophilicity is the ratio of water area to nitrogen area multiplied by one hundred.

$$\%\ \text{hydrophilicity} = 100\ \frac{S_{water}}{S_{nitrogen}}$$

The corresponding hydrophobicity is:

% hydrophobicity = 100 - % hydrophilicity

Hydrophilicity values for some organic and inorganic pigments are listed in Table 1.3. The strongest hydrophobicity is shown here by phthalocyanine blue. However, for the sake of completeness, it must be pointed out that in this pigment class hydrophobicity also depends on the crystal modification. In the present case we are dealing with an acicular β-modification. As would be expected, iron (III) oxide and titanium dioxide, due to the large difference between the electron affinity of their atoms, have the strongest polarity and consequently the highest hydrophilicity in this list.

Another method for characterising pigments with regard to their polarity is based on the fact that the wetting of a solid by a liquid is an exothermic process. It is possible to measure the amount of heat released during the wetting process by means of a sensitive calorimeter. Thus by relating this amount of heat to the unit area of the pigment surface, a characteristic value for the corresponding pigment/solvent-combination can be obtained: the **specific enthalpy of wetting** ΔH, which is a measure for the interaction between the pigment surface and the surrounding solvent [15].

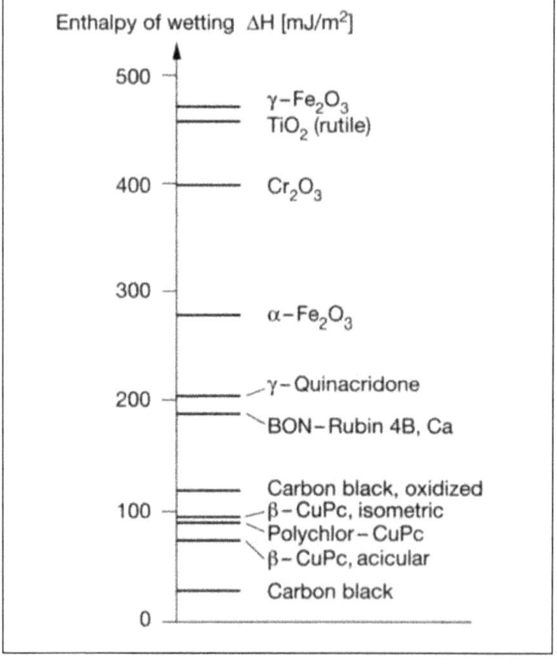

Figure 1.11: Polarity scale of pigment surfaces derived from the enthalpies of immersion according to Sappok

Using values for the specific enthalpy of wetting obtained with water as the wetting medium, *Sappok* [16] drew up a polarity scale for some organic and inorganic pigments (Figure 1.11).

The polar inorganic pigments, with their high surface energy, are found in the upper part of the scale; the weakly

Table 1.3: Degree of hydrophilicity of some organic and inorganic pigments

Pigment	Nitrogen area S_{N_2} [m²/g]	Water area S_{H_2O} [m²/g]	Degree of hydrophilicity [%]
β-Cu-phthalocyanine	70	2.5	3.6
γ-Quinacridone	61	8.2	13.4
γ-Iron (III) oxide	15	10.1	67.3
Titanium dioxide (rutile)	5	4.0	80.0

polar and non-polar organic pigments, in contrast, are positioned in the lower part. The slight influence of particle shape on the polarity of copper phthalocyanine blue is easy to discern.

1.4.2 Surface tension and capillarity

1.4.2.1 Surface tension

Thermodynamic approach

Molecules at the surface of a phase (solid or liquid) have a special position in comparison with those located within the bulk, as they are not subjected to equal forces in all directions. Whereas forces acting on molecules in the interior are mutually compensated vectorially, the molecules forming the surface lack outer neighbours, whereby they experience an inwardly directed force. This force is termed **dynamic surface tension**. Consequently, work must be performed against the surface tension in order to bring a molecule from the bulk to the surface. This gives rise to the fact that surface molecules possess a certain amount of free potential energy. The total free potential energy of all surface molecules is designated as the **free surface energy**. It makes itself felt as a force acting upon the whole length of the surface and striving to oppose any increase of it. Since this force manifests itself by an inward pulling effect, it can also be interpreted as a negative pressure.

The free surface energy F_0 is a function of pressure p, temperature T and the surface properties. Thus, the surface A also appears as a variable state quantity in the thermodynamic interpretation of surface tension, i.e.:

$$F_0 = \mathrm{f}\,(p, T, A)$$

The total differential of the free surface energy is given by the relation:

$$\mathrm{d}F_0 = \left(\frac{\partial F_0}{\partial p}\right)_{T,A} \mathrm{d}p + \left(\frac{\partial F_0}{\partial T}\right)_{P,A} \mathrm{d}T + \left(\frac{\partial F_0}{\partial A}\right)_{P,A} \mathrm{d}A$$

Should the change of state expressed by this relation take place at constant pressure and temperature, the enthalpy rather than the energy must be considered. The change in the free surface enthalpy with increasing surface then reduces to:

$$dG_0 = \left(\frac{\partial G_0}{\partial A}\right)_{p,T} dA$$

Equation 1.6:

$$\frac{dG_0}{dA} = \left(\frac{\partial G_0}{\partial A}\right)_{p,T} = \sigma$$

The symbol σ represents the **surface tension**. According to Equation 1.6, it has to be interpreted as the (reversible) work required to increase the interface one unit area under isothermal and isobaric conditions.

The concepts surface tension and interfacial tension

Strictly speaking, the term surface tension should be used only when substances (solid or liquid) in vacuum are considered. This is a fictitious state that can seldom be realised experimentally. In practice, the surfaces of solids and liquids stand under the influence of a gaseous phase (in most cases, air), so that surface tensions measured under such conditions are in reality **interfacial tensions** between the solid or liquid and surrounding gaseous phase. However, it should be noted that the effect of molecules of gases at very low pressure on the surface molecules of a solid or liquid can generally be neglected. If the surface tension of a liquid is measured, e.g. by the drop method, first in air and then in a chlorine atmosphere – all other conditions being kept constant – different values would result. For this reason, some authors have proposed the use of the term **interfacial tension** and revert to the denomination **surface tension** only in exceptional cases, e.g. in the case of liquids in equilibrium with their own vapour. In the present monograph we shall speak of **surface tension** (with symbol σ) in connection with liquids in air and of **interfacial tension** (with symbol γ) in respect of solid/solid, solid/liquid and liquid/liquid interfaces. Most of the relationships concerning surface tension explained here are valid for interfacial tension as well.

Interfacial tension and wetting

The surface tension of a solid, i.e. also of a pigment, is closely connected with its wettability. Wetting occurs at the interface between a solid and a liquid if the adhesive forces between the molecules of solid and liquid are stronger than the cohesive forces between the atoms or molecules of the liquid. Since wetting is

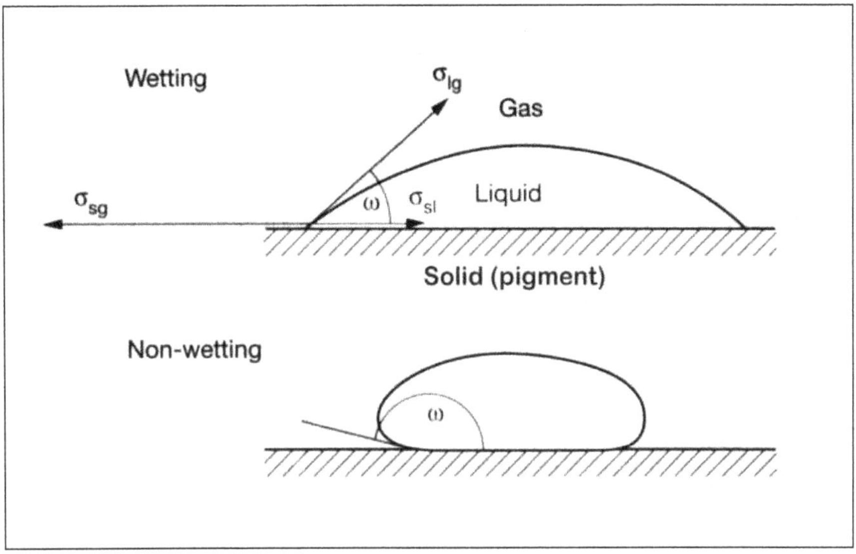

Figure 1.12: Wetting relationships on pigment surfaces
σ_{lg}: surface tension of the liquid, σ_{sg}: surface tension of the solid
σ_{sl}: surface tension at the interface (interfacial tension)
ω: contact angle

the result of an interaction, it is not a specific property of the liquid but is also dependent on the composition of the solid. Thus, a particular liquid may wet one solid but not another.

Measurement of liquid surface tension

The surface tension of liquids can be determined directly by experimental methods. These are based on different interfacial phenomena such as drop formation when a liquid falls from a narrow tube, the generation of bubbles by conducting a gas through a liquid, the contact angle resulting at the solid/liquid interface, the formation of a liquid film after dipping and detaching a ring of wire from a liquid, or the elevation or depression of a liquid in a capillary tube. The methods are classified as **dynamic** or **static**, depending on whether the interface is altered during the measurement or not. They are discussed very fully by *Dörfler*[17].

Angle of contact, wetting tension and spreading

The surface tension of solids cannot be as easily determined as is the case with liquids, and it is therefore necessary to describe the wetting processes at the solid/liquid interface using indirect methods. Two quantities play the decisive role here: the **contact angle** and the liquid/gas interfacial tension. The relationship can be determined as follows: If a drop of a liquid is placed onto the smooth surface of a solid (e.g. pigment), the drop will acquire different shapes depending on the nature of this solid. Figure 1.12 illustrates the shapes of a wetting and a non-wetting liquid on the same solid.

At the junction of the three phases, solid, liquid and gaseous, the so-called contact angle ω is formed. At this point, the three interfacial tensions *solid/gas, solid/liquid* and *gas/liquid* are in equilibrium. Being vectors, their relationship can be described according to the parallelogram rule by the following expression known as the **Young equation**:

Equation 1.7:

$$\vec{\sigma}_{sg} = \vec{\sigma}_{sl} + \vec{\sigma}_{lg} \cos \omega$$

We can conclude from this Equation 1.7 that the contact angle depends only on the participating substances. The difference $\sigma_{sg} - \sigma_{sl}$, denoted **wetting tension**, refers to the work that is reversibly obtained when the solid/liquid interface increases by 1 cm^2. When the wetting tension assumes negative values (i.e. when $\cos \omega < 0$), it is also termed **adhesive tension**.

Depending on the magnitude of the contact angle, two different cases may be distinguished:

1. The contact angle is smaller than 90° ($\cos \omega > 0$). The wetting tension is positive; the liquid wets the solid surface. In the case of $\omega = 0°$ we have complete wetting. The liquid does not form a drop, but spreads across the surface as a thin layer. This phenomenon is called spreading.
2. The contact angle is larger than 90° ($\cos \omega < 0$). The wetting tension is negative. In this case, only partial wetting or no wetting at all occurs. The drop lies slightly flattened on the solid surface.

Thus, carbon tetrachloride (dipole moment = 0 Debye[1]) and water (dipole moment = 1.85 Debye) placed on the

Table 1.4: Contact angles of fluid/solid combinations

System	Contact angle ω [°]
Water/paraffin	≈ 105
Water/glass	0
Mercury/glass	140
Benzene/glass	6
Water/PTFE	108
Benzene/PTFE	46
Carbon tetrachloride/ Cu-phthalocyanine	0
Ethylene glycol/Cu-phthalocyanine	54
Water/Cu-phthalocyanine	89

[1] Debye = $3.33564 \cdot 10^{-30}$ Cm
For the sake of simplicity, we write σ_l, σ_s and γ_{sl} instead of σ_{lg}, σ_{sg} and σ_{sl} from Equation 1.7, respectively

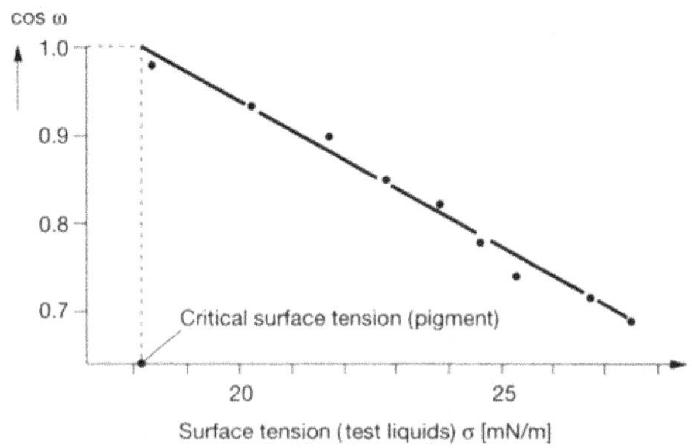

Figure 1.13: Example for determination of the critical surface tension according to Zisman

non-polar copper phthalocyanine blue give contact angles of 0°, i.e. complete, and 89°, i.e. only partial, wetting respectively (see also Table 1.4).

The reason for the different behaviour of liquids towards other phases lies in the intermolecular cohesive forces, which also take effect on the interfaces. These forces consist primarily of van der Waals forces. Hydrogen bonding forces may also play a part, e.g. when in contact with water. The stronger the intermolecular forces, the weaker the interfacial tension resulting between two phases. Table 1.4 gives contact angle data for some solid/liquid systems.

Critical surface tension

A method to ascertain whether a given liquid wets or even spreads on a solid surface was developed by *Zisman*. For this purpose, a series of (preferentially homologous) liquids is placed on the solid test surface and the resulting contact angles are measured. By plotting the cosines of these contact angles against the surface tensions of the respective liquids, an approximate straight line is obtained (Figure 1.13). Extrapolation to a zero contact angle gives a value that is characteristic for the solid in question and was termed the **critical surface tension** by *Zisman*. This parameter corresponds to a theoretical liquid that is capable of completely wetting the test solid and beginning to spread on it. A smooth surface is a prerequisite for this procedure, as rough surfaces tend to favour wetting [18].

Since knowledge of the surface tension of a liquid by itself is not enough to judge if the liquid is suitable for complete wetting of a given solid, the critical surface tension is a valuable supplement for settling this question. When the liquid surface tension is lower than the critical surface tension of the solid, it follows immediately that the liquid must spread on the solid. More details on this topic may be found in the pertinent literature [19, 20].

Solid surface tension

According to *Young*, the interfacial tension between two liquids or a liquid and a solid ought to be approximately equal to the difference between the surface tensions of the substances involved[1]. Thus, the wetting tension at the solid/liquid interface (sl) under the condition $\sigma_s < \sigma_l$ should be:

Equation 1.8:

$$\gamma_{sl} = \sigma_l - \sigma_s$$

This is attributed to a type of compensation effect arising from a mutual attraction between the surface molecules of the participating substances across the interface. Hence, only the non-compensated part of the higher surface tension remains effective. Due to the impossibility of determining σ_s by direct methods, this theory has not been experimentally proven to date, and Equation 1.8 remained controversial for a long time until *Antonov* (1942) showed its validity at least for the case of saturated miscible liquids. Thus, this relationship is generally referred to in the literature as **Antonov's rule.**

Assuming that the forces nearly compensate one another at the **solid/liquid** interface, a useful relationship for the determination of solid surface tension can be stated. Its deduction was performed as follows:

If the gaseous phase in Equation 1.7 is air, Young's equation can be written as:

$$\sigma_s = \gamma_{sl} + \sigma_l \cos \omega$$

In this equation, γ_{sl} can be replaced by $\sigma_l - \sigma_s$ from Equation 1.8

$$\sigma_s = \sigma_l - \sigma_s + \sigma_l \cos \omega$$

and thus:

Equation 1.9

$$\sigma_s = \left(\frac{1}{2}\right) \sigma_l \, (1 + \cos \omega)$$

As σ_l and ω can be measured directly, the problem of assessing the surface tension of a solid ends with the measurement of the contact angle which a liquid with known surface tension forms on the solid. On the other hand, if the surface tension of a solid is known, it is possible to quantitatively predict the contact angle with the help of Equation 1.9 and thus the wetting of any liquid of known surface tension.

The technical realisation of contact angle and surface tension measurements on organic pigments was obtained indirectly by high vacuum deposition of the

[1] For the sake of simplicity, we write σ_l, σ_s and γ_{sl} instead of σ_{lg}, σ_{sg} and σ_{sl} from Equation 1.7, respectively.

Table 1.5: Values of interfacial tension in mN/m for pigment/solvent combinations

Pigment C.I. Denomination and N°	Water σ_f = 72.8 mN/m	Glycerol σ_f = 63.4 mN/m	Ethylene glycol σ_f = 47.7 mN/m
Pigment Violet 23 51319	30.6	31.1	31.5
Pigment Blue 15:3 74160	35.0	34.2	34.0
Pigment Yellow 1 11680	36.0	33.6	35.6
Pigment Red 3 12120	35.0	32.9	33.2
Pigment Red 122 73915	35.9	32.7	35.2

pigments onto suitable substrates [21–23]. The pigment layers thus produced made feasible the measurement of contact angles formed by a series of liquids. Some of these results are compiled in Table 1.5.

Surface tension and temperature

The temperature dependence of surface tension is also of interest for the pigment processor, as the pigment dispersing process may frequently attain temperatures of about 70 °C. As a rule, the surface tension of liquids decreases linearly with increasing temperature, the contact angle diminishes and hence wetting becomes correspondingly better.

1.4.2.2 Capillarity

The presence of interstices such as pores, crevasses and capillaries in pigment agglomerates requires knowledge of the connections between surface tension and the phenomena denoted **capillarity** for a better understanding of the wetting process. As is generally known, liquids in narrow tubes stand higher or lower than would be expected according to the law of communicating vessels. Wetting liquids are raised, non-wetting ones are depressed. The height difference is caused by an upward or downward force that is a consequence of surface tension. Thus, a wetting liquid will rise up the tube till the upward force $2\,\pi\,r\,\sigma$ and the weight of the liquid column $\pi\,r^2\,h\,\rho\,g$ counterbalance and equilibrium occurs. In these terms, r is the capillary radius, σ the surface tension of the liquid, h the height of the liquid column, ρ the liquid density and g the gravity constant. (The slight curvature of the liquid surface has been neglected). Equating the two opposing forces:

$$2\,\pi\,r\,\sigma = \pi\,r^2\,h\,\rho\,g$$

yields

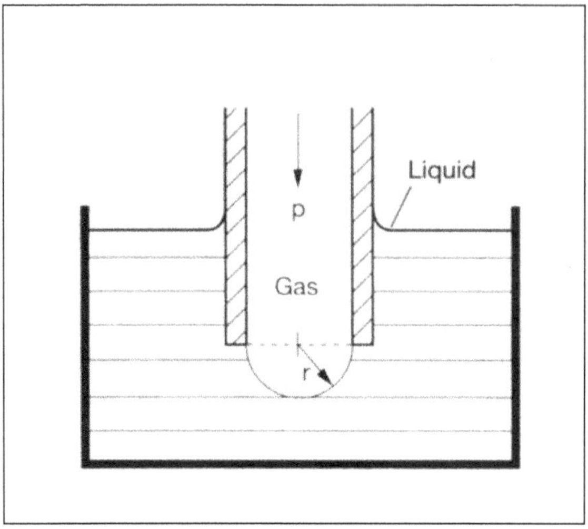

Figure 1.14: Emergence of a gas bubble from a capillary tube due to pressure. p = pressure, r = radius of the bubble (For derivation of capillary pressure)

$$h = \frac{2\sigma}{r \rho g}$$

When the liquid does not completely wet the capillary wall, this equation must be modified by multiplying the fraction on the right-hand side by cos ω. This should be taken into consideration particularly for precision measurements.

A further important relationship exists between capillary pressure (to be defined in Equation 1.10) and surface tension. It is the mathematical connection between these two quantities for the case of curved surfaces, as expressed by the *Gauss-Laplace* equation. For this purpose, let us examine a bubble formed by overpressure (e.g. by blowing) at the end of a capillary tube immersed in a liquid (Figure 1.14). The pressure difference between the gas pressure inside the bubble and the pressure exerted by the liquid on the bubble is called **capillary pressure** p_k. For spherical bubbles ($F = 4\pi r^2$ and $V = \frac{4}{3} \pi r^3$), the differential change of surface energy with the differential change of the radius dr is equal to $8 \pi \sigma r \, dr$. The volumetric work inside the bubble, i.e. the work required to form the bubble, is $p_k V$ and its differential is $p_k dV = p_k 4 \pi r^2 dr$. At the point of equilibrium both will be equal:

$$p_k 4 \pi r^2 dr = 8 \pi \sigma r \, dr$$

Therefore it follows that the capillary pressure:

Equation 1.10:

$$p_k = \frac{2\sigma}{r}$$

If the bubble is non-spherical, then l/r is replaced by the arithmetical average of the reciprocal values of the main radii r_1 and r_2 of curvature. Correspondingly, the capillary pressure in the Gauß-Laplace fundamental equation becomes:

Equation 1.11:

$$p_k = 2\sigma \frac{1}{2}\left(\frac{1}{r_1}+\frac{1}{r_2}\right) = \sigma\left(\frac{1}{r_1}+\frac{1}{r_2}\right)$$

i.e. directly proportional to the surface tension as in Equation 1.10. It is easily seen that for the special case of a spherical bubble $r_1 = r_2$ and Equation 1.11 simplifies to Equation 1.10.

As indicated, the capillary pressure is, as a rule, the driving force for the penetration of a liquid medium in the pores and capillaries of pigment agglomerates. These relationships will be discussed in more detail in the Section 4.2.2 *Wetting.*

1.4.3 Electrical polarisability

1.4.3.1 Polarisation of dielectrics in an electric field

Some properties of pigment molecules, such as refractive index and the interaction with surfactants or macromolecules, are directly related to their electrical polarisability. The interaction with components of the medium has, on the other hand, a decisive effect on the extent of adsorption and the degree of stability of the pigment dispersion.

As is generally known, matter is composed of atoms, ions and molecules. Atoms consist of electrons and a nucleus. In contrast to conductors, the electrons in non-conductors (also called **insulators** or **dielectrics**) can only move inside the atoms or molecules. If a dielectric is introduced into an electric field, the field will cause a slight displacement of the electric charges in the dielectric molecules. A new field thus results inside the dielectric slightly acting against the external field, which appears then weakened in the interior of the dielectric. This phenomenon is called **electrical polarisation.**

According to *Debye's* theory of molecular polarisation (in the year 1912), the reduction of the external field can be attributed to two causes: On the one hand, we have separation of the electrical centres of positive and negative charge in the dielectric molecules due to the action of the external field, whereby the atoms or molecules become dipoles. The dipole charges are, however, mutually compensated within the dielectric, whereas the charges at the surfaces remain effective. The polarised state can therefore only be ascertained at the surfaces (Figure 1.15). This kind of polarisation is termed **induced** or **distortion polarisation**. It occurs in all dielectrics and is not influenced by temperature.

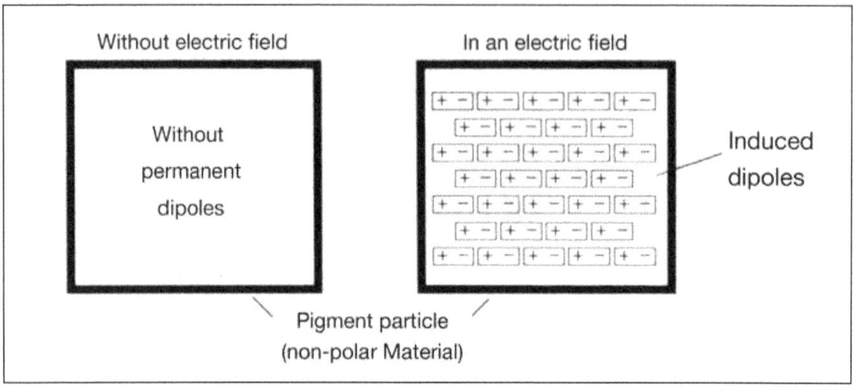

Figure 1.15: Idealised representation of distortion polarisation in a (non-polar) pigment by induced displacement of charges in an electric field)

On the other hand, the dielectric may already contain permanent dipoles, i.e. its molecules are polar. When not submitted to an electric field, these permanent dipoles are oriented at random due to the thermal motions of the molecules. No outward polar effect is observed. However, as soon as an external electric field is applied, the dipoles undergo an orientation by preferentially aligning themselves in the direction of the field (Figure 1.16). This **orientation polarisation**, as it is denoted, is temperature-dependent because it is counteracted by the molecular thermal motion.

Consequently, the *total* polarisation is the sum of *induced* and *orientation* polarisation.

According to electrostatic laws, the induced dipole moment p_i in each atom or molecule is proportional to the field strength E^* of the local field, i.e. the electric field actually acting on the atoms or molecules.

$$p_i = \alpha \, E^*$$

The proportionality factor α is a measure of the ease with which charge separation in a molecule is induced by an external electric field. It is called **molecular polarisability** and is a characteristic quantity for the atoms or molecules of a dielectric.

1.4.3.2 Polarisability and refractive index

Being an electromagnetic wave radiation, light is also in a position to produce polarisation of electrical charges in the lattice structure of a pigment crystal. The dipoles thus induced in ions or molecules, just like permanent dipoles already present, oscillate at the same frequency as the exciting electromagnetic radiation

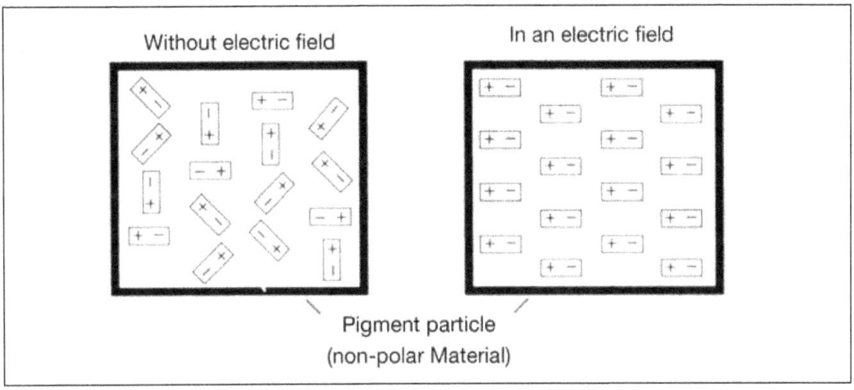

Figure 1.16: Orientation polarisation in a (dipolar) pigment in an electric field (The permanent dipoles are orientated at random in the absence of an electric field

thereby emitting secondary waves. These interact with the exciting radiation, and the resulting wave thus generally travels with reduced light velocity in the pigment crystal.

Since the absolute refractive index is the ratio of the speed of light in a vacuum to the speed of light in the medium considered, then a change in the refractive index with polarisation is the logical consequence. The refractive index of a pigment rises with increasing polarisability of its structure units. In other words, the refractive index can therefore serve as a measure of the substance's electrical polarisability.

1.4.3.3 Polarisability and pigment/binder-interaction

As will be shown in more detail, the colloidal stability of pigment dispersions is based on the adsorption of macromolecules at the particle surfaces. The adsorbed binder layer either shields the particle from the attractive forces of neighbouring particles or gives rise, by virtue of its ionic groups (e.g. in the case of polyelectrolytes), to a repulsion between particles carrying the same electric charge. Sometimes both mechanisms may contribute to the stabilisation of a system. (More information on this topic is given in Section 4.2.3, **Stabilisation**.)

The ionic portions of these macromolecules adsorbed at the pigment surface are not only bound by Coulombic forces. The polarisability of both pigment molecules and adsorbed ionic groups has, as a consequence of the dipole moment depending on the polarized state, also a decisive influence on the accommodation and immobilisation processes. This rise in the adsorption capacity of the particle surface is therefore directly proportional to the molecular polarisability and to the volume of the adsorbed ionic groups, as their polarisability is increased by a larger volume.

1.4.4 Surface treatment

The section entitled **surface character** would be incomplete if lacking a discussion on surface treatment. Surface-treated pigments contain one or several substances that cover their surface either by physisorption or by chemical bonding. This coverage, also called **coating**, may consist of a monomolecular layer or also of a multilayer.

Surface preparation

Pigments are often coated with inorganic and organic substances. Inorganic surface treatments change the nature of the pigment surface so that, in a colloid chemical sense, the coated pigment is changed into the substance that it is covered with. Whereas inorganic surface treatments remain on the pigment particles, organic surface treatments, which are just adsorbed to the surface, tend to detach when the pigments are put into formulations. Inorganic surface treatments can improve flocculation stabilization whereas organic surface treatments enhance dispersibility and can improve wetting efficiency. Such products are known as **readily dispersible pigments**.

A good example is titanium dioxide, a pigment that may be dispersed very easily in aqueous systems on account of the polar character of its surface, but not so in organic media. By covering the particle surface with trimethylchlorsilane, a chemical reaction with the hydroxyl groups of the hydrated pigment takes place. The methyl groups of the $-Si(CH_3)_3$-units are aligned outwardly and confer an organophilic character to the surface. A pigment treated in this manner is readily dispersible in organic media [24].

Apart from dispersibility and flocculation stability, further application properties like hiding power, transparency, tinting strength and binder absorption can be optimised by surface treatment. Thus, many surface-treated pigments have improved flow properties in comparison with their non-treated counterparts. This is attributed to the fact that by pre-wetting the surface with a coating agent, the binder absorption is diminished, resulting in a lower increase in viscosity of the system. On the other hand, by coating pigment particles with inorganic compounds like SiO_2 and Al_2O_3, it is possible to considerably improve the heat fastness of some inorganic pigments, e.g. iron oxide yellow. Furthermore, carbon blacks pre-treated with oxygen-containing groups show, depending on the binder system, better dispersibility, flow and optical properties. Owing to their low electrical conductivity, these carbon-black grades are particularly appropriate for colouring systems requiring good insulating properties [25].

Organic pretreatment

The coating of azo pigments with rosin during their precipitation has been known since the 1920s. As a rule, the rosin is deposited unto the pigment as a free abietic acid or as a soluble alkaline-metal salt. This treatment impairs the crystal growth, whereby fine-grained products with good dispersibility and high transparency and tinting strength are obtained. Important pigments for printing inks, such as the Litholrubin grades, are treated in this manner.

Another method consists of adding aliphatic amines to the pigment suspension, either as salts in aqueous solution or as free amines in aqueous emulsions. In the first case, the conversion of the amine salt into a free amine is brought about by the addition of an alkali before the pigment filtration takes place. The amine bond to the pigment surface is especially strong, and for this reason it is assumed that the bond is chemisorptive in nature. By virtue of the outwardly-oriented hydrocarbon residue, amines impart an organophilic character to the pigment.

Coating the pigment surface with suitable polyvinyl chloride-copolymers results in 'tailor-made' products for the coloration of PVC. Trials have been carried out for the same purpose in which the polyvinyl chloride-copolymer was replaced by plasticiser of the phthalic acid ester type, but the results were unsatisfactory.

Complications

It goes without saying that the surface treatment of pigments also gives rise to some disadvantages. Their field of application is strongly limited, as the substance used for coating must be compatible with the medium to be coloured. The coating agent may also have an unfavourable effect on certain properties of the pigmented system. Amine-treated pigments, for example, have a tendency to impair the drying process of coatings hardening by oxidation. Despite the disadvantages in the surface treatment of pigments, its significance is increasing, and it has become an essential factor and, above all, a simple tool in order to obtain products with optimised application properties.

Further information on this topic is supplied in the relevant literature [26–28].

2 Optical properties of pigmented systems

When rays of light strike a pigmented layer (coating, print or coloured plastic), the interaction taking place between the layer and light results in reflection, scattering and absorption of the incident radiation (Figure 2.1). The overall optical impression gained by an observer depends on how and the extent to which the pigmented layer influences and governs these three phenomena. Decisive factors for the interaction are:

1. The optical constants of the pigment and the medium surrounding it, i.e. refractive index, absorption and scattering coefficients.
2. The pigment particle size.
3. The wavelength of the incident light.
4. The chemical constitution of the pigment.
5. The concentration of the pigment in the system.

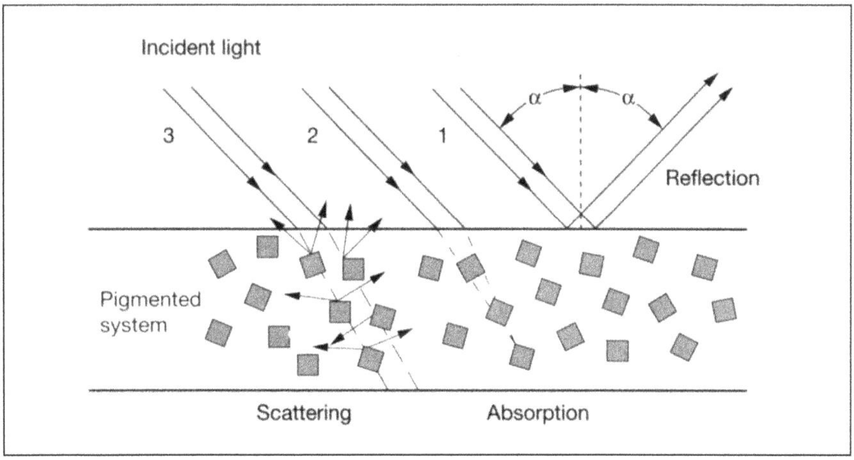

Figure 2.1: Interaction between a light beam and a pigmented layer. Ray 1 strikes the interface at an angle α and is reflected. Ray 2 penetrates into the layer changing its direction and is absorbed by the pigment particles. Ray 3, also refracted like Ray 2 (possibly at different wavelength), is scattered at the pigment particles. About 4 % of the incident light intensity results in reflection, the remaining 96 % in absorption and scattering

Juan M. Oyarzún: Pigment Processing

ISBN 978-3-86630-665-3

The substrate on which the pigmented system is applied can also have an effect on the optical behaviour, unless the portion of light striking it is so insignificant that the layer can be considered as hiding.

2.1 Reflection, gloss and gloss haze

2.1.1 Gloss

The fraction of the incident radiation that is reflected at the surface of the pigmented layer is responsible for the gloss. It can be affected by the refractive index of the system, the angle of incidence, the radiation intensity and the surface structure (rough or smooth), which determines the spatial distribution of the reflected light. If the surface is absolutely smooth (an ideal case for pigmented systems), it reflects light in only one direction, i.e. at an angle equal to the angle of incidence. This is known as **regular** or **specular reflection**. Such a surface exhibits gloss – in most cases even high gloss. When the surface is not smooth, the incident radiation will be reflected or scattered in all directions, that is we are dealing with a **diffuse reflection.** Such surfaces are perceived as matt or, at best, as semi-glossy. The mechanisms of reflection are shown in Figure 2.2.

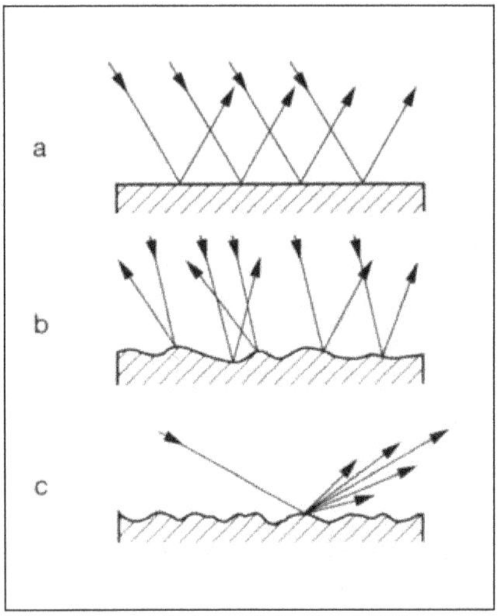

Figure 2.2: Types of reflection on surfaces: specular reflection on a smooth surface (a), diffuse reflection on a rough surface (b) scatter reflection caused by microscopic surface irregularities (c)

Gloss in pigmented systems is also affected by the degree of dispersion and the particle size distribution of the pigments present. These two factors play an important role, particularly in the case of extremely thin layers, e.g. prints. Should the surface be disturbed by protruding agglomerates, this gives rise to a reduction in gloss. In those cases where the reflecting layer shows a strong absorption, its absorption coefficient also has an influence on the gloss. Further details concerning the influence of pigments on gloss may be found in the articles by *Zorll* and *Braun* [29, 141, 142].

As is the case with the sensation of colour, gloss perception is also the sum of different effects of a physical, optical-physiological and psychological nature. Only the purely physical processes are measurable, and these follow the laws of geometric optics and depend on the surface structure of the pigmented layer and substrate as well as on the kind of radiation and its intensity.

Conditions for light reflection

The phenomenon we term reflection is based on the interaction between light waves and the atoms or molecules lying near the surface of the reflecting medium. These atoms or molecules are stimulated by the alternating electromagnetic field of the light wave, and emit a secondary radiation of the same frequency as the incident radiation. In addition, the secondary radiation is partly linearly polarised. The human eye, however, cannot distinguish between normal and polarised light. The reflectivity, i.e. the ratio of the intensity of the reflected radiation to that of the incident radiation, is affected by the wavelength, the relative refractive indices of the media involved and the angle of incidence. As the first medium is air, the relative refractive indices are practically identical with the absolute values. The relationship between the degree of reflectivity and the factors influencing it is given by the well-known Fresnel formulae, which in turn can be derived from Maxwell's well-known equations for describing the processes in changing electromagnetic fields.

A schematic picture of the situation is shown in Figure 2.3. When the angle of incidence is oblique, the electric field of the incident waves can be divided into two components. The first component, or electric vector, oscillates perpendicular,

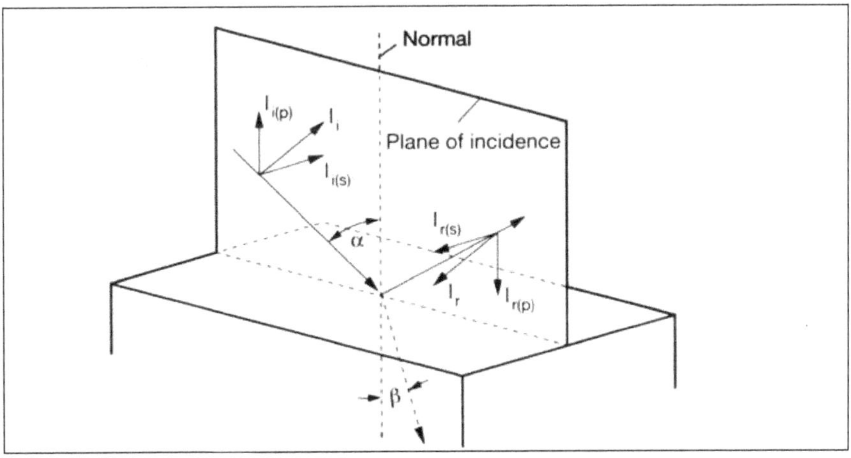

Figure 2.3: Division of the light vectors for derivation of the Fresnel equations: α angle of incidence; β angle of refraction; I_i intensity of the incident radiation; I_r intensity of the reflected radiation

the second parallel to the plane of incidence. If the radiation intensity is equal in both components, i.e.:

Equation 2.12:

$$I_{i(s)} = I_{i(p)} = \frac{I_i}{2}$$

the corresponding formulae for the reflectivity are:

$$R_p = \left(\frac{I_r}{I_i}\right)_p = \frac{\tan^2(\alpha-\beta)}{\tan^2(\alpha+\beta)}$$

$$R_s = \left(\frac{I_r}{I_i}\right)_s = \frac{\sin^2(\alpha-\beta)}{\sin^2(\alpha+\beta)}$$

where:

α = the angle of incidence
β = the angle of refraction (i.e. the angle formed between the refracted beam and the normal)
I_i = the intensity of the incident radiation,
I_r = the intensity of the reflected radiation,
R_s = the reflectivity of the light polarised perpendicular to the incident plane,
R_p = the reflectivity of the light polarised parallel to the incident plane.

Therefore, the reflectivity R_1 at *oblique* incidence of light is:

$$R_1 = \frac{I_r}{I_i} = \frac{I_{r(p)} + I_{r(s)}}{I_i}$$

Taking Equation 2.12 into account, we obtain Equation 2.13:

Equation 2.13:

$$R_1 = \frac{1}{2}\left(R_p + R_s\right)$$

At normal incidence ($\alpha = 0$) there is no distinction between both oscillating components, because the plane of incidence is not defined. Thus, if the reflectivity at *normal* incidence is termed R_2 we obtain:

$$R_2 = R_s = R_p = \frac{\sin^2(\alpha-\beta)}{\sin^2(\alpha+\beta)}$$

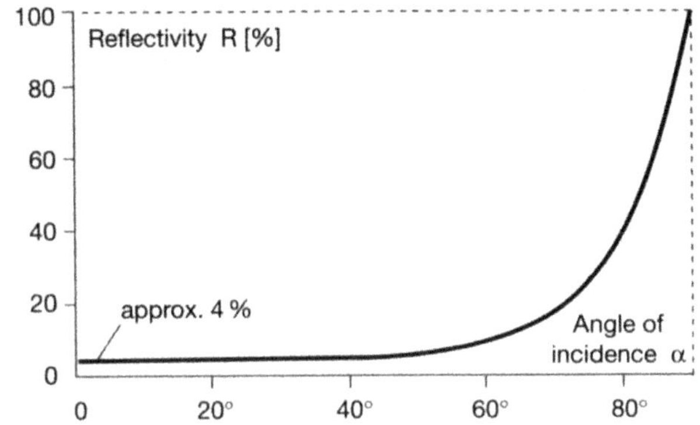

Figure 2.4: Reflectivity as a function of angle of incidence

Using the relevant trigonometric relationships, furthermore Snell's refraction law

$$n = \frac{\sin\alpha}{\sin\beta}$$

and finally the fact that the cosines of very small angles are practically 1, we derive Equation 2.14:

Equation 2.14:

$$R_2 = \left(\frac{n\sin\beta - \sin\beta}{n\sin\beta + \sin\beta}\right)^2 = \left(\frac{n-1}{n+1}\right)^2$$

For most of the polymer binders employed in pigment processing, as well as for glass, the value of n is approximately 1.5. Thus, in the particular case of normal incidence, about 4 % of the incident energy is reflected at the surface.

Fresnel's relations are not only valid for the visible light, but also for UV and infrared radiation and for any angle of incidence. The dependence of the reflectivity on the angle of incidence for a medium with a refractive index of 1.5 is shown in Figure 2.4. The R_I-values were calculated with the aid of Equation 2.13, and the corresponding angles of refraction in accordance with Snell's law. It may be easily seen that the reflectivity remains practically constant at 4 % up to an angle of incidence of 35°, and that it shows no significant increase until 55° is reached.

Gloss measurement

As the results are strongly dependent on the angle of incidence, it is important to choose the most suitable measuring geometry. Three distinct angles of incidence have been laid down, covering the whole range of gloss from absolutely matt to high-gloss surfaces, i.e. 20° for high-gloss surfaces, 60° for surfaces of moderate gloss and 85° for matt surfaces. The incident angles of 45° and 75° were also standardized, and are used in the ceramics, plastics and paper industries [30].

Special instruments termed reflectometers or goniophotometers are used for gloss measurement. In reflectometers, the intensity of the specular reflection is determined in an angular range limited by diaphragms. The criterion for the gloss is the "reflectometer value". The gloss-dependent distribution of scattered light is practically disregarded in this method. The standard consists of a high-gloss, black glass plate, and its reflected radiation, due to the complete surface evenness, passes through the diaphragm aperture without loss. Provided that this glass plate has a refractive index of 1.567, the standardised procedures assign it a reflectometer value of 100. The gloss of the sample is thus expressed as a percentage of the standard. Figure 2.5 illustrates the path of radiation in a **reflectometer**.

The **goniophotometer** can measure both the intensity of the specular reflection and the spatial distribution of the scattered radiation, thus allowing the evaluation of radiation reflected at any desired angle for angles of incidence set between 0° and

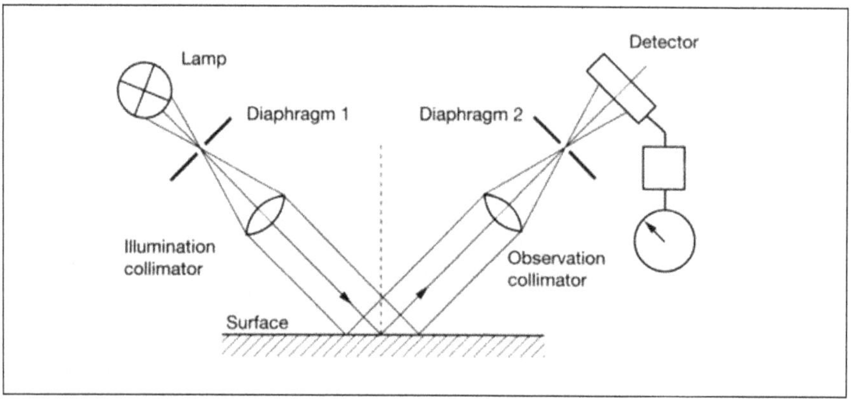

Figure 2.5: Radiation path in a reflectometer. Using an optical system (diaphragm and collimator for illumination), the light coming from the lamp is reflected at the surface and passes (fully or partly, depending on the gloss grade) through the aperture of a second diaphragm. The intensity of the reflected radiation is thus a measure for the gloss of the surface, registered by a photoelectric detector. (A more versatile device, the goniophotometer, is based on the same principle, however, with the illumination-collimator fixed at selected angles of incidence, and with the observation-collimator's angular position variable in nearly the whole quadrant).

75°. Assessment of the intensity of the spatial distribution can thus be carried out for different directions of illumination. The plot of light intensity against the angle of observation generally gives rise to a bell-shaped curve (Figure 2.6). The curve maximum as well as the half-value width can be taken as criteria for the gloss.

2.1.2 Gloss haze

When viewing the image of a very bright, punctiform light source in the high-gloss surface of a pigmented layer, it is occasionally possible to observe that the image is not absolutely distinct, but is surrounded by a halo. This is perceived as a milky haze in the case of dark shades. This phenomenon, known as **gloss haze** or **reflection haze**, is caused by microscopic surface irregularities in the size of 2 to 20 µm giving rise to scattered light of low intensity. In other words, the major portion of the radiation is reflected specularly (mirror-like) and the rest is scattered close to the angle of reflection (Figure 2.2c). The surface irregularities are attributed to pigment agglomerates or flocculates in the surface region of the layer [31].

The human eye is particularly sensitive to the decrease in distinctness-of-image on high-gloss surfaces caused by gloss haze, and in such cases the reflectometer value alone is insufficient for gloss characterisation. To alleviate this problem, gloss-meters (with an angle of illumination of 20°) were developed having, in addition to the reflectometer diaphragms, two detectors in the angular ranges of 17.3° to 19.1° and 20.9° to 22.7° for measurement of the diffuse radiation caused by microscopic surface roughness. The optical system of such gloss-meters is shown in Figure 2.7 [32, 33].

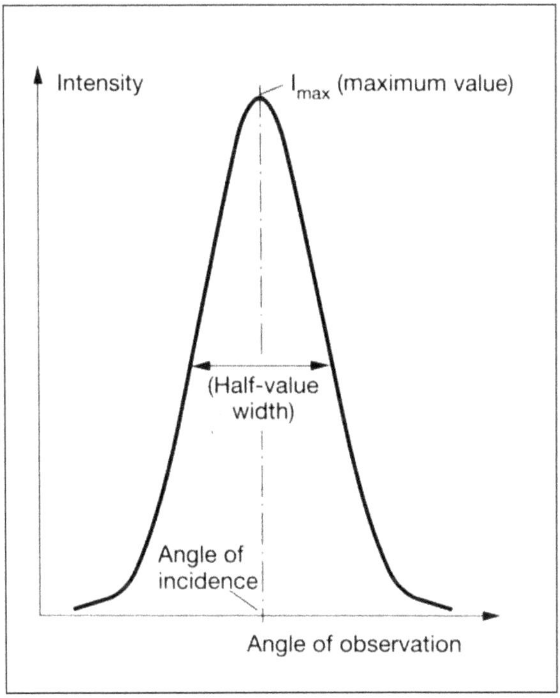

Figure 2.6: Goniophotometric curves of a glossy surface: Light intensity as a function of the viewing angle. Characterisation by maximum intensity and half-value width

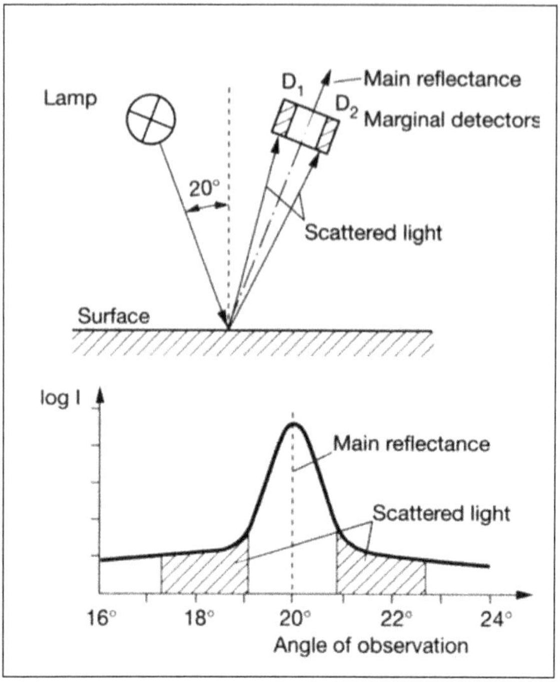

Figure 2.7: Determination of gloss haze. Measurement of the intensity of scattered light I beside the main reflection at the angle zones of 17.3° to 19.1° and 20.9° to 22.7° by two additional detectors D_1 and D_2

Due to the correlation between gloss or gloss haze and particle size in pigmented layers, it is possible to resort to gloss development as a criterion for the degree of dispersion of pigments. As gloss continues to increase during dispersion – even after the tinting strength has attained a practically constant value –, gloss can be, in principle, a more accurate measure of the degree of dispersion. An appropriate method is described in DIN EN ISO 8781 with introduction in DIN EN ISO 8780 (see also Section 5.3.2).

A standard for the determination of gloss haze on paint films is DIN EN ISO 13803. The main standards for gloss measurement are DIN EN ISO 2813 and ASTM D 523.

2.2 Light scattering by colloidal disperse particles

The remainder of the radiation that is not reflected at the surface of the pigmented layer enters the layer itself and undergoes a change in direction and velocity (Figure 2.1). In addition to this change, known as **refraction**, the light waves are, depending on their wavelength, partly **absorbed** or **scattered** by the pigment particles beneath the surface. If the layer is not thick enough to hide it, the rest of the light reaches the substrate where it is reflected diffusely. This section will deal mainly with the optical processes in hiding systems.

Both optical phenomena originating within the pigmented layer, absorption and scattering, are decisive for the colour and hiding power, i.e. the ability of the system to eliminate the colour or colour differences of the substrate (DIN EN ISO 18451-1).

It was mentioned at the beginning of this section that light scattering is based on an interaction between light waves and the irradiated medium. When light impinges on matter, atoms or molecules are excited and they emit in their turn light waves (termed secondary radiation) which have the same frequency as the primary radiation. If the medium is not homogeneous, as is the case with pigmented systems, the secondary radiation is scattered. However, such inhomogeneity is not an essential prerequisite. Light scattering may also occur in a chemically homogeneous substance if statistical fluctuations in the molecular distribution gave rise to the formation of slightly more dense regions, where the refractive indices differ from that of their surroundings.

When the scattering particles become so small that their surface does not allow the formation of a uniform, plane wave-front, the laws of geometrical optics are no longer valid. The scattering processes occurring at such particles can then only be explained by the interpretation of light and its interaction with matter as a wave phenomenon.

2.2.1 Rayleigh scattering

The physical mechanism of light scattering by small particles was studied by, among other investigators, *W. Rayleigh* and *G. Mie*. When the scattering is caused by particles much smaller than the wavelengths of the incident light, it is termed **Rayleigh scattering**. The scattering by particles with sizes comparable with, or larger than the wavelengths of visible light is the subject of the **Mie theory**. A good means for distinguishing between the kinds of light scattering provides the **Mie scattering parameter** q, which is equal to $2\pi r/\lambda_M$, where λ_M is the wavelength in the medium of refractive index n_M ($\lambda_M = \lambda/n_M$) and r the radius of the scattering particles. When q for scattering, but non-absorbing, particles is found to be < 1, we are dealing with Rayleigh scattering. If $1 < q < 2$, the scattering corresponds to a transitional phase from Rayleigh to Mie scattering. Rayleigh scattering of natural, unpolarised light at air molecules or colloidal disperse particles is brought about in practice by particles with sizes up to 50 nm.

Wavelength dependence of Rayleigh scattering

The quantitative approach to Rayleigh scattering requires two conditions:

1. The primary radiation induces a dipole in each particle. These dipoles are then the source of the scattered radiation. However, the field strength of the primary

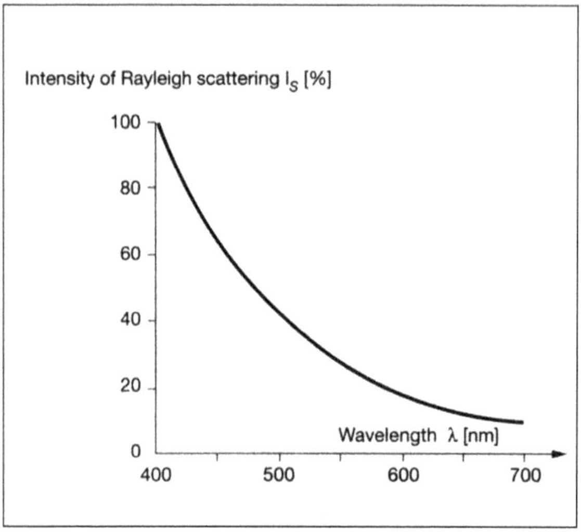

Figure 2.8: Intensity of Rayleigh scattering as a function of wavelength

radiation must be uniform everywhere, which is only the case when the scattering particles are very small in comparison with the wavelength.

2. The particles must be far enough apart to exclude possible interactions between them.

When these prerequisites are fulfilled, the dependence of the intensity of the scattered light I_s on the factors that affect it is given by the Rayleigh equation:

Equation 2.15:

$$I_s = k I_0 \frac{V^2}{\lambda^4}$$

I_0 is the intensity of the primary radiation, λ its wavelength, V the volume of the particle and k a constant that includes the distance of the scattering source to the detector and the interaction between light and the scattering particles in the form of polarisability. This relationship shows that the intensity of the secondary radiation is proportional to the intensity of the primary radiation and to the inverse fourth power of the wavelength. Hence, the short wavelength fraction (blue, violet) of white sunlight is more strongly scattered than the long (red) one, an evident explanation for the blue appearance of a clear sky. The intensity of scattered light as a function of wavelength is shown in Figure 2.8.

Rayleigh scattering as a function of the angle of observation

Taking the further assumption that the incident radiation is unpolarised, the dependence of the intensity of the secondary radiation on the angle of observation φ (also called scattering angle), i.e. the angle between the incident and the scattered beam, is given by:

$$I_{s\varphi} = \frac{1}{2} I_{max} (1 + \cos^2 \varphi)$$

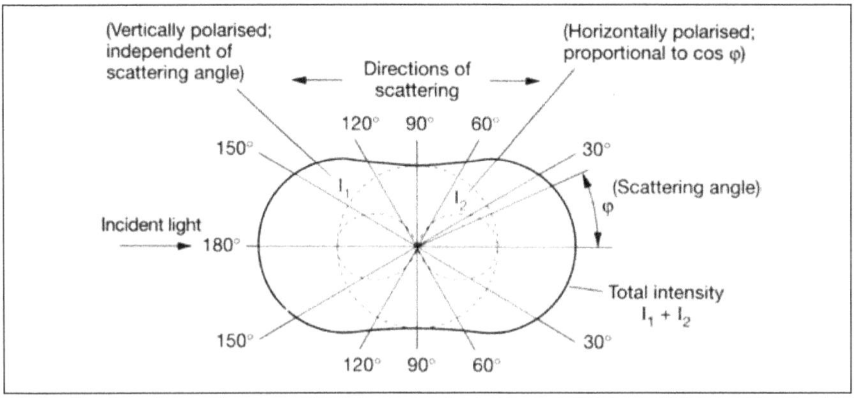

Figure 2.9: Polar diagram of the angular dependence of the intensity of Rayleigh scattering

In this case the scattered radiation consists of two components, I_1 and I_2, polarised at right angles to one another. The first component, vertically polarised, is **independent of the scattering angle**. The second one, horizontally polarised, is **proportional** to $\cos^2 \varphi$. Their addition gives the total intensity distribution of the scattered radiation. This is plotted against the scattering angle in Figure 2.9.

The scattering intensity attains its maximum value at the angles 0° and 180°. The intensities are equal in the forward and backward direction relative to the direction of propagation of the primary radiation. According to the extended Rayleigh theory, this symmetrical scatter pattern is also valid for all colloidal particles, provided that both conditions previously mentioned are fulfilled.

2.2.2 Mie theory

When the scattering elements are no longer small in comparison with the wavelength, the intensity of the scattered radiation depends in a fairly complicated manner on wavelength, scattering angle, particle size and shape. In physical and technical literature this kind of scattering is referred to as **Mie scattering** [34].

A detailed discussion of Mie's theory would go beyond the scope of this monograph, and we would refer the reader to a book by *van de Hulst* [35]. We shall concern ourselves here with the basic aspects of the theory which are needed for a better understanding of the optical properties of pigmented systems.

Interference theory and Mie effect

The chosen model consists of spherical isotropic particles with complex refractive index n^* (including the k) and absorption index k, being struck by a wave front.

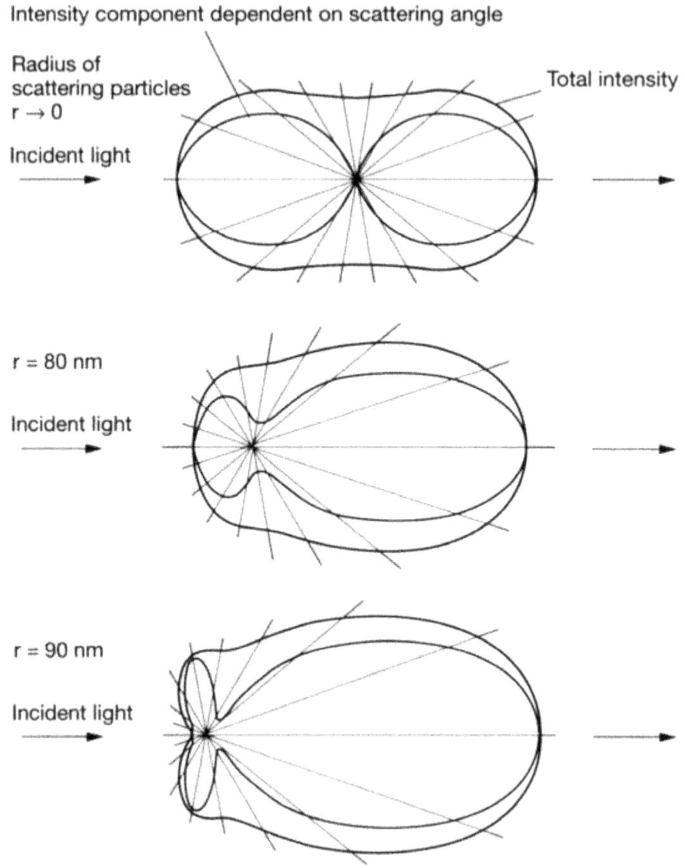

Figure 2.10: Polar diagrams for intensity distribution as a function of the angle of observation by Mie scattering in colloidal metal solutions with differing sizes of scattering particles. The principles of the Mie scattering theory also apply to pigmented systems, e.g. a pigmented paint, thus leading to similar diagrams. (See text for further explanations)

In addition, it is assumed that the particles are separated from each other by distances large enough to exclude any interaction between them. The actual problem is to find the solution of Maxwell's field equations applicable to this particular model, and for this purpose a system of spherical polar co-ordinates originating at the particle centre is introduced. Moreover, we also bring in the scattering parameter q already mentioned (Section 2.2.1, page 61).

As in the Rayleigh scattering case, the *Mie* scattered radiation also consists of two components polarised at right angles to each other. The initial values for calculating these components are the scattering parameter, the refractive indices of particles and medium and the scattering angle. The model conditions eliminate interference of the scattered radiations with one another, so that the total energy of the scattered light is equal to the sum of the light energy scattered by the individual particles. This results in the scatter diagrams, conveniently represented in polar co-ordinates, in Figure 2.10.

Extremely small particles give a scatter diagram characteristic for Rayleigh scattering, i.e. the intensities of the scattered light are equal in the forward and reverse directions. The diagram is thus symmetrical about a plane through the centre of the particle and perpendicular to the direction of incident light. The length of the radius vector is proportional to the intensity in the corresponding direction (Figure 2.10 above).

When the dimensions of the particles are comparable to the wavelength, i.e. when they attain the transitional stage between Rayleigh and Mie scattering, emission centres are induced at different locations on the same particle, giving rise to dipoles and multi-poles (quadrupoles, octapoles) emitting scattered waves in different phases. These "wave packets" may, depending on the scattering direction and geometrical conditions, interfere with one another, resulting in an attenuation of the scattered radiation which is proportional to the scattering angle. In the polar diagrams of the intensity distribution, the direction of propagation of the exciting radiation becomes the preferential direction of the scattered radiation. Thus, with reference to total scattering, forward scattering increases and backward scattering decreases. This phenomenon is termed the **Mie effect**. In the case of even larger particles, maxima and minima appear in the intensity distribution and their position is characteristic for the size of the scattering particles (Figure 2.10 centre and below). The deviation from Rayleigh scattering may be evaluated by the **dissymmetry ratio** Z, which is the ratio of the scattering intensities at $\varphi = 45°$ and $\varphi = 135°$.

It is customary to denote the size x of a round scattering particle with the diameter by the quotient between its circumference and the wavelength of the incident light $x = \pi d / \lambda$. As a rule of the thumb, Raleigh scattering occurs when x is smaller than 0.1. When x is very much larger than 1, Fresnel reflection takes place, whereas, when x is in the range of about 1, scattering is well described by the Mie theory.

Extinction. Lambert's law

Light radiation is absorbed and scattered during its passage through a pigmented medium. The resulting attenuation of the radiation intensity, denoted as **extinction**, is proportional to the incident intensity I_0 and the path length x in the medium. The differential decrease, dI, per path element dx is given by the relationship:

$$\mathrm{d}I = -\varepsilon I \mathrm{d}x$$

The proportionality factor ε is termed the extinction coefficient. Integrating between the limits I_0 and I, the equation known as Lambert's law is obtained:

Equation 2.16:

$$I = I_0 e^{-\varepsilon x}$$

It should be noted that I_0 is the radiation actually entering the pigmented medium, i.e. after correction for any loss (e.g. reflection) at the surface. In a simultaneously absorbing and scattering medium, the extinction coefficient is composed additively of the absorption coefficient K and the scattering coefficient S. Taking this into account, Equation 2.16 becomes:

$$I = I_0 \cdot e^{-(K+S)x}$$

K and S are specific constants for the system, depending solely on wavelength. They describe the attenuation of the radiation per unit path and have thus the dimension of reciprocal length. When $K + S = 1/x$, the exponent of e in Equation 2.16 becomes equal to -1. Hence, from the purely physical point of view, the extinction coefficient can be considered as that path, at which the intensity of radiation decreases by an e-th portion. This is also valid for absorption or scattering alone, if in the extinction process either the scattering or the absorption respectively can be neglected.

Quantitative approach to absorption and scattering

Extinction processes may be quantitatively described with the aid of the **absorption** and **scattering cross sections** Q_a and Q_s, which are the cross-sections of the radiation beam affected by absorption and scattering respectively. These two quantities do not coincide with the geometrical cross-section of the particle, the reason for this lying in the fact that light scattering occurs due to the interaction of the electrical field of the light and the electrons in a dielectric substance. The ratios of absorption and scattering cross-sections to the geometrical cross-section are termed **absorption** and **scattering efficiency** F_a, and F_s, or also **efficiency factors for absorption and scattering**.

$$F_a = \frac{Q_a}{\pi r^2} \qquad F_s = \frac{Q_s}{\pi r^2}$$

In these equations, r is the particle radius. The factors F_a and F_s are dimensionless. If we denote the number of particles per unit volume as N, the relationships between absorption or scattering coefficients and the corresponding effective cross-sections are:

Equation 2.17a: $K = N Q_a = N F_a \pi r^2$

Equation 2.17b: $S = N Q_s = N F_s \pi r^2$

For particles of any size and refractive index, the absorption and scattering coefficients depend upon wavelength, size, refractive index and angle of observation. In the case of particles that simultaneously *absorb* and *scatter* light, their normal refractive index n becomes a complex number n^* ($n^* = n - ik$), whose imaginary term includes the absorption index. In general, the dependence may be expressed mathematically by:

$$F = f(\lambda, r, n, k, \varphi)$$

For non-absorbing particles, $F_a = 0$ and for non-scattering particles, $F_s = 0$. Taking the assumption that the particles are spherical, the volume of an individual particle is $(4/3)\,\pi\, r^3$. Thus we obtain for the particle volume concentration c_V:

$$c_V = N \frac{4}{3} \pi r^3 \text{ ; and hence } \qquad N = \frac{3}{4} \frac{c_V}{\pi r^3}$$

However, according to the *Mie* scattering parameter, $r = q\, \lambda_M / 2\pi$. Using these expressions for N and r, Equation 2.17b may be rewritten:

$$S = \frac{3}{2} c_V \frac{\pi}{\lambda_M} \frac{F_s}{q}$$

And, for the absorption coefficient in Equation 2.17a:

$$K = \frac{3}{2} c_V \frac{\pi}{\lambda_M} \frac{F_a}{q}$$

As a general rule, the result is represented as the scattering or absorption efficiency in dependence of the Mie scattering parameter. In the first case this type of plot results in the curve of a damped oscillation. Its shape for particles with n = 1.80 is given in Figure 2.11.

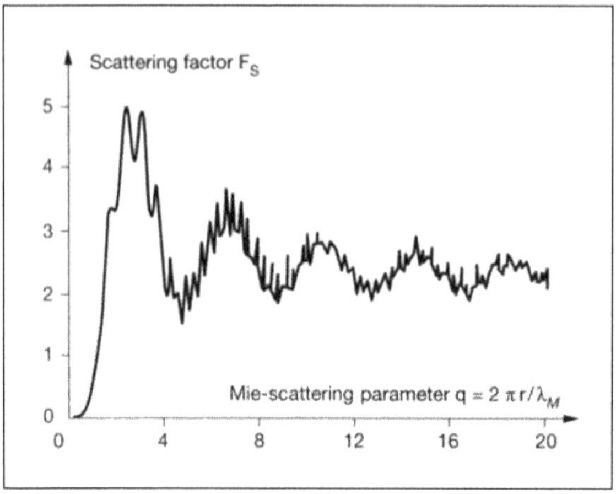

Figure 2.11: Mie scattering: plot of scattering factor F_s versus Mie parameter q (r radius of the scattering particles, λ_M wavelength in the medium

Source: G. Benzing et al. Pigmente und Farbstoffe für die Lackindustrie, Expert Verlag, Grafenau, 1992

The curve shows a series of maxima and minima, the so-called **interference structure**. When the scattering particles are much larger than the wavelength, the scattering efficiency becomes independent of wavelength. This is one of the essential differences between Rayleigh and Mie scattering. It can be seen that for larger values of q, the scattering cross-section becomes twice the geometrical cross-section. This may be attributed to additional extinction by the diffraction of light at the edges of the geometrical cross-section of the particles. This shape of curve also applies for other refractive indices [36].

Pigment relevant consequences from the Mie theory

Thus, the main conclusions to be drawn from the Mie theory are:

As regards absorption

- When the particles are extremely finely dispersed, further comminution does not result in an increase of absorption.
- The absorption of very small particles increases with increasing absorption index.
- With particles having a high absorption index and a low refractive index (e.g. organic pigments), the optimum size lies below the limit of measurement.
- Particles with low absorption index and high refractive index (e.g. coloured inorganic pigments) have a definite absorption maximum.

As regards scattering

- There is an optimum particle size that depends on wavelength.
- For large particles, in comparison with the wavelength, light scattering is independent of wavelength.

- The scattering cross-section of large particles, again compared with wavelength, is twice the geometrical cross-section.
- For a given particle size, there exists a specific wavelength at which light scattering attains a maximum value.

Since the Mie theory takes more factors into account, it is more general than Rayleigh scattering which may actually be considered to be a special case of Mie scattering.

In fact, the very idealised conditions assumed by the *Mie* theory do not occur in practice for pigment applications. Pigment particles are neither absolutely spherical nor are they of a uniform size. Nevertheless, this theory allows a satisfactory interpretation of the optical behaviour of pigmented systems, as we shall see in the next section.

2.3 Absorption and scattering as factors in the tinctorial properties of pigments

The tinctorial properties of pigments, such as hiding and lightening power, tinting strength, saturation, transparency, semi-transparency and hue (shade), are primarily governed by the absorption and scattering power of the pigments used. Further factors are the refractive indices of pigment and medium, particle size and shape, particle size distribution and pigment volume concentration.

In the following, we will discuss in detail how these factors affect the coloristic properties of pigmented systems.

2.3.1 Hiding power, semi-transparency and transparency

The hiding power concept

From a purely physical point of view, pigments do not have a hiding power because it is always the pigmented print or coating that hides the substrate. However, for practical reasons it was considered convenient to attribute a hiding power to the pigments themselves. This is feasible, provided that comparisons for characterisation purposes are carried out under identical conditions, i.e. medium, dispersing method, concentration, layer thickness, kind of illumination etc. Thus, if coatings are produced with a series of pigments under equal and defined application and measuring conditions, differences in their hiding power can be attributed to the pigments. Under these circumstances the concept of **pigment hiding power** is by no means inadmissible. The pigment technologist must merely be fully aware of what is actually meant by this notion.

The role of scattering and absorption

The influence of scattering and absorption coefficients upon hiding power and transparency may – depending on the kind of pigment – differ considerably. White pigments have a high scattering coefficient and an almost negligible absorption coefficient. Their hiding power is consequently based on their considerable ability to scatter light. Black pigments, on the other hand, exhibit a high absorption coefficient and a low scattering power throughout the visible spectrum. Hence, they hide by virtue of their high absorption. Coloured pigments can be hiding, semi-transparent or transparent, depending on which property, absorbing or scattering, gains the upper hand. With few exceptions, the hiding power of organic pigments results from their high absorption coefficient (e.g. Cu-phthalocyanine blue or dioxazine violet). For coloured inorganic pigments, where the refractive indices are generally much higher, hiding power is mainly governed by their light-scattering power.

Pigments that hide mainly due to their high absorption are called *semi-transparent*. The concept of **semi-transparency** in this context consequently comprises **low scattering** combined with **high absorption**. Therefore, transparent colourings are possible with semi-transparent pigments if these are applied in low concentrations. Unfortunately, the technical literature often does not differentiate between the two concepts of transparency and semi-transparency. In contrast to semi-transparent pigments, opaque colourings with transparent pigments (e.g. transparent iron oxides) are not possible.

Relationship between refractive index and absorption

In Section 2.2, light scattering was explained as secondary radiation emitted by oscillating dipoles in an irradiated pigment particle. This secondary radiation is superimposed on the surrounding wave field and is partly extinguished. The larger the difference between the refractive indices of pigment and binder, the larger is that part of the secondary radiation which is not extinguished. Hence, pigments with a high refractive index contribute more strongly to hiding power than pigments having a low refractive index. At the same pigment concentration, systems containing rutile titanium dioxide ($n = 2.75$) thus have a higher hiding power than systems with anatase titanium dioxide ($n = 2.53$).

The refractive index of anisotropic pigments is not only dependent on the wavelength, but also on the direction of the crystal axis. This fact is taken into account in practice by employing the average value of the various refractive indices in the crystal.

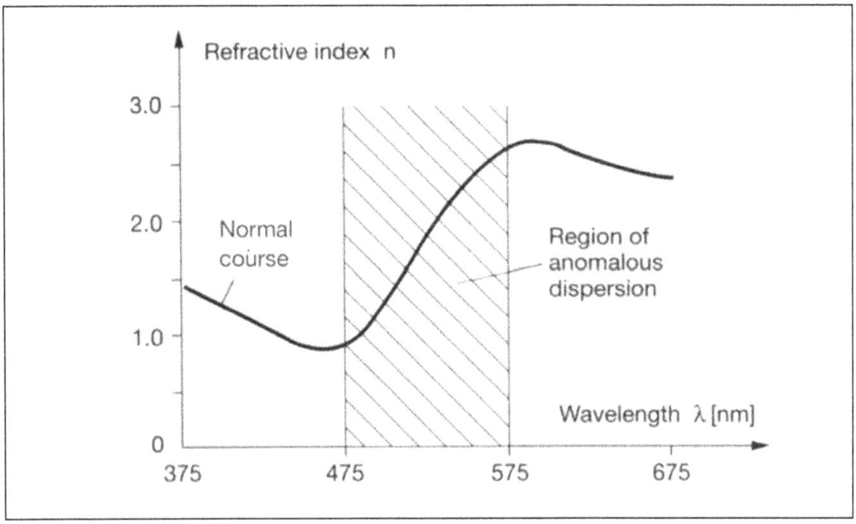

Figure 2.12: Anomalous dispersion curve of solid fuchsine

Source: L. Bergman and C. Schaefer Experimentalphysik [137]

The speed of propagation of light waves through matter is, contrary to its speed in vacuum, dependent on wavelength (or frequency). This phenomenon is known as **dispersion**[1]. The **dispersion curve**, as the plot of refractive index versus wavelength or frequency is termed, of non-chromatic substances merely shows a continuous increase of the refractive index with decreasing wavelength ($dn/d\lambda < 0$). This case is termed a **normal dispersion**. The refractive index of white pigments is considered to be constant throughout the visible spectrum. Strictly speaking, this is not true. Most of the white pigments absorb in the UV-range, which implies a slight absorption in the violet/blue-region.

Coloured pigments behave differently in this respect. The absorption of substances that absorb selectively takes place together with a gradual change in refractive index, which decreases on the shorter wavelength side of the absorption band and increases on the longer wavelength side ($dn/d\lambda > 0$). The dispersion curve takes a normal course in the neighbouring zones. In this case we are dealing with an **anomalous dispersion**. The higher the absorption, i.e. the absorption coefficient, the stronger the decrease in n on the shorter wavelength side. By way of example, Figure 2.12 illustrates the anomalous dispersion of solid fuchsine. The hatched surface corresponds to the zone of high absorption.

[1] Note the different meaning of "dispersion", with respect to the pigment's optical behaviour and to the distribution of pigment in a medium.

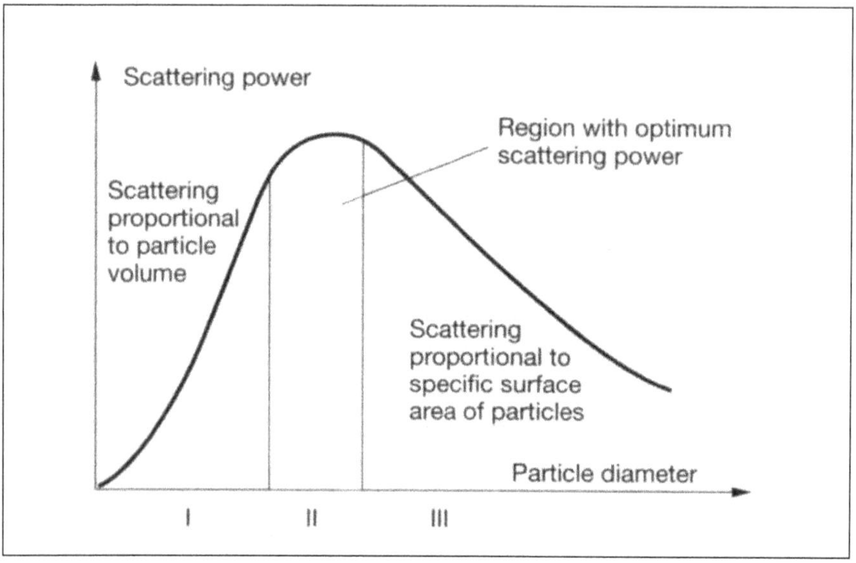

Figure 2.13: Scattering power as a function of particle diameter at constant wavelength and given refractive indices of pigment and medium (schematic)

The explanation for the anomalous dispersion in light-absorbing substances is found in the electromagnetic nature of light waves. Light velocity, and thus the refractive index, is determined by the dipoles which the incident light induces in the pigment or dyestuff. The induced dipole oscillates with an **'eigen'-frequency** that lies in the UV and visible regions. At this 'eigen'-frequency, the highest energy transfer between the exciting electromagnetic wave and the dipole takes place, and is tantamount to an absorption maximum. Coloured pigments scatter only in a wavelength region in which the refractive index attains its maximum value.

Particle size and scattering

As mentioned in the preceding section, light scattering is also a function of particle size. In this context, an important conclusion of the Mie theory states that there is an optimum particle size for scattering. The relationship between particle diameter and scattering power at a constant wavelength and given refractive indices of pigment and medium is represented schematically in Figure 2.13.

Scattering by particles which are extremely small in relation to wavelength is very slight, and occurs as Rayleigh scattering (Region I). With increasing particle radius, scattering initially increases rapidly but then, during the transition stage from *Rayleigh* to Mie scattering, it rises more and more slowly as a consequence of the increasing Mie effect. Scattering attains a maximum value (Region II) at a particle size characteristic for every pigment which, since it lies in the magnitude of the half-wavelength of incident light, corresponds to the critical particle radius. Beyond this maximum, scattering decreases proportionally to the specific surface area. Regular scattering takes place in this region (Region III). In practice, the optimum particle size giving maximum scattering is not an absolute value, as it is governed not only by the pigment but also by its interaction with the medium in which it is embedded.

To date, an exact calculation of the optimum particle diameter for scattering is not possible. The following relationship, which was found empirically by *H. H. Weber*[38], enables good estimates to be made:

Equation 2.18:

$$D^* \approx \frac{\lambda_M}{2.1\left(n_p - n_B\right)}$$

where λ_M is the wavelength at maximum scattering, n_P the refractive index of the pigment and n_B that of the binder. For a rutile titanium dioxide with $n_P = 2.75$ embedded in a binder with $n_B = 1.5$, Equation 2.18 results in an optimum particle diameter of approximately 210 nm at $\lambda = 550$ nm.

It has already been mentioned that titanium dioxide exhibits a weak absorption in the short wavelength range ($\lambda < 430$ nm), giving rise to a slight yellow tone of the pigmented material. Pigment producers counteract this effect by choosing an average particle size which lies somewhat below the optimum. Thus, scattering in the short wavelength range, i.e. of the blue radiation, is increased.

It follows from the Mie theory that the optimum particles sizes for hiding power and for tinting strength do not coincide in the case of coloured pigments. The pigment producer must therefore either find a compromise or optimise his pigment with regard to one of these properties at the expense of the other. Due to their high refractive indices, inorganic pigments are generally prepared for good hiding power, whereas the dominant criterion for organic pigments, because of their high absorption, is the tinting strength. About in the eighties of the last century, special grades of organic pigments optimised with regard to scattering, and hence hiding power, came onto the market. Their shades make them of interest for the coatings industry as possible substitutes for lead chromates and lead molybdates. Moreover, only transparent and semi-transparent pigments can be considered for transparent colourings and four-colour printing. Light scattering is unwelcome in these applications.

Scattering power and pigment volume concentration

A further factor influencing the intensity of light scattering is the volume of pigment. For this reason, when the hiding power of a pigmented system is being determined, it is convenient to express the pigment concentration in terms of volume concentration. The **pigment volume concentration** (PVC) is the ratio of pigment volume to total volume multiplied by 100.

Equation 2.19:

$$\text{PVC} = \frac{\text{pigment volume}}{\text{pigment volume} + \text{medium volume}} \; 100 \, [\%]$$

In dry coatings, the volume of the medium is equal to the sum of the volumes of all non-volatile constituents except pigment. Time and again we find the remark in technical literature that to calculate the PVC of systems containing extender, the extender volume must be added to that of the pigment. When the connection between hiding power and PVC is under discussion, this would be incorrect, because the extender does not contribute to the scattering power of the system.

The dependence of the hiding power of a white pigment (expressed in m^2/l of the dry coating) on the pigment volume concentration is illustrated in Figure 2.14. At low PVC values, the hiding power increases linearly with the concentration. However, with rising concentration the optical interaction between the pigment particles becomes larger. As a consequence, the increase in hiding power slows down and attains a maximum. If the PVC rises beyond this maximum, "shadow" effects resulting from the particles' close proximity reduce the optically effective pigment volume and hence the hiding power.

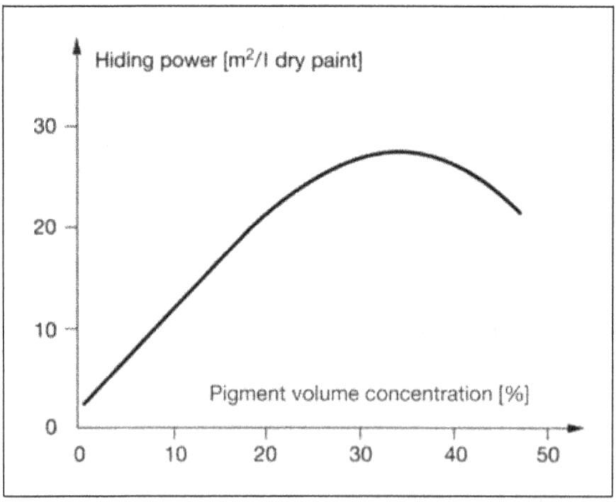

Figure 2.14: Hiding power of a white pigment in dependence on pigment volume concentration

Source: Kronos Guide [24]

Hiding power and particle shape

Taking the Mie theory as a basis, *Gans* investigated the influence of particle shape on light scattering. He came to the conclusion that anisometry favours an increase in hiding power

[39]. Subsequent testing with acicular and spherical zinc oxide particles showed a lower hiding power for the latter. A similar result was found with zinc sulphide and some organic lakes.

Determination of scattering and absorption coefficients

The Kubelka-Munk theory [40–42] results in a relatively easy method for assessing the absorption and scattering coefficients of a given pigment. The fundamental Kubelka-Munk equation (20):

Equation 2.20:

$$K/S = (1 - R_\infty)^2/2\, R_\infty$$

stating the relationship between the absorption and scattering coefficients K and S and the reflectance R_∞ of an opaque ("infinitely thick") coating layer, is also valid for mixtures of several pigments. In this case, the absorption and scattering coefficients of the blend are additive functions of the absorption and scattering coefficients of the individual pigments present, multiplied by their concentrations (c) in the blend. Thus, the Kubelka-Munk equation for such a pigment blend is:

$$\frac{K}{S} = \frac{c_1K_1 + c_2K_2 + c_3K_3 + \cdots + c_nK_n}{c_1S_1 + c_2S_2 + c_3S_3 + \cdots + c_nS_n}$$

The reflectance functions, F, for a white pigment (W) and a coloured pigment (B) are given by the following relationships:

$$F_W = \frac{K_W}{S_W} \qquad F_B = \frac{K_B}{S_B}$$

For a blend consisting of 1 part of white pigment and c parts of coloured pigment, its reflection function F_M is:

Equation 2.21:

$$F_M = \frac{K_M}{S_M} = \frac{K_W + cK_B}{S_W + cS_B}$$

If the scattering coefficient of the white pigment is arbitrarily given the value 1, then $K_W = F_W$ and Equation 2.21 becomes:

$$F_M = \frac{F_W + cK_B}{1 + cS_B}$$

Replacement of the term S_B in this equation by its value K_B/F_B and solving for K_B yields:

Equation 2.22:

$$K_B = \frac{F_B(F_W - F_M)}{c(F_M - F_B)}$$

With the aid of the relationship $S_B = K_B/F_B$, the corresponding scattering coefficients are finally obtained. By way of example, Figure 2.15 shows the plot of K_B versus wavelength for a dioxazine pigment.

The method requires three coatings; one with the white pigment, another with the coloured pigment and a third with a coloured/white reduction. However, since the reflectance of the white pigment is generally known from earlier tests, two coatings are often sufficient in practice. Using modern data-processing equipment, calculations at, for the sake of accuracy, thirty or more different wavelengths is no longer a problem. Standard methods for assessment of the relative values of *S* and *K* are specified in DIN EN ISO 787-24 and DIN 53 164.

Determination of hiding power

In order to determine the hiding power of a pigmented material, a number of coatings (or foils) of various thicknesses are applied over a black and white substrate (contrast or hiding-power chart). That coating or foil thickness is then determined at which the colour difference DE (white/black) as specified by DIN 55987 is 1. The reciprocal of this thickness is the hiding power, and states how many square meters can be coated with one litre of *dry coating* according to the chosen opacity criterion.

A different criterion is specified by DIN EN ISO 6504-1 and ASTM D 2805. Here a coating is considered as hiding when the ratio Y_s/Y_w is higher than or equal to 0.98. Y_s and Y_w are the CIE[1] tristimulus values

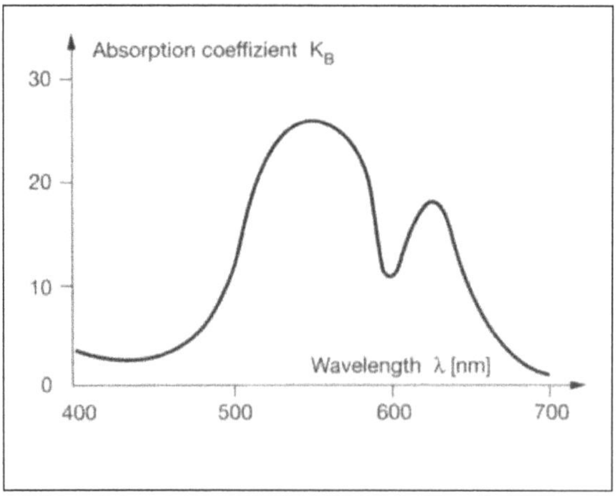

Figure 2.15: Wavelength dependence of the absorption coefficient of dioxazine violet

on a black and white undercoat respectively. A method for the visual determination of hiding power of dried paints using a wedge-shaped layer is described in DIN 55601.

Assessment of transparency

The optical properties transparency and semi-transparency discussed at the beginning of this section have low scattering in common. Transparency (or semi-transparency) with pigmented materials is only possible when the pigment particles are significantly smaller than half the wavelength of the impinging radiation and therefore insufficient surface area is available for scattering. Thus, transparent iron oxides have particles between 50 and 100 nm long, and 10 and 20 nm wide; the particles of transparent Prussian-blue types are smaller than 20 nm and those of transparent cobalt blue are between 20 and 100 nm.

As specified in DIN 53988, transparency (or semi-transparency) is determined by applying a layer of the pigmented material over a black substrate. The smaller the colour difference between sample and substrate, the better the transparency (semi-transparency) of the system. It is expressed by the dimension T, termed **transparency number.** This is the quotient of the layer thickness or concentration h and the colour difference to the ideal black substrate ΔE^*_{ab} (i.e. CIE tristimulus values X, Y, Z equal to 0), calculated according to DIN EN ISO 11664-4:

Equation 2.23:

$$T = \frac{h}{\Delta E^*_{ab}(h)}$$

However, the transparency number of a coating system is, generally speaking, not a constant, but is dependent on h. An exact assessment of this dependence requires two determinations. The transparency number for the difference in film thickness or concentration h existing between the specimens P_1 and P_2 may be calculated using the relationship:

$$T = \frac{h}{\frac{h-h_1}{h_2-h_1}\left(\Delta E_2 - \Delta E_1\right) + \Delta E_1}$$

In this expression, $\Delta E_i = \Delta E^*_{ab}$ is the colour difference observed with the system over a black substrate when compared with the ideal black. (According to DIN EN

[1] CIE: Commission Internationale de l'Eclairage (International Commission on Illumination)

ISO 11664-4, the colour difference ΔE is the vector sum of three components, i.e. lightness difference ΔL and chromacity differences Δa and Δb).

h_i is the film thickness or concentration of the relevant specimens;

h is the film thickness in mm or the pigment mass (or volume) relative to area in g/m^2 (or l/m^2) between h_1 and h_2.

If the transparency numbers of P_1 and P_2 are equal, Equation 2.23 can be applied.

2.3.2 Lightening power

A further optical property of white pigments that is closely associated with their scattering power is the **lightening power**. Such is the term for the ability of a white pigment to raise the lightness of a coloured, grey or black material (DIN 55982). Lightening power characterises the yield of a white pigment and corresponds to the tinting strength of coloured pigments. Hiding and lightening power are the most important properties of white pigments.

With increasing scattering power of the white pigment, the average path-length of light in the system becomes shorter and, consequently, the absorption of the coloured or black pigment decreases. This leads to a decrease in the **saturation**, i.e. the proportion of pure spectral colour, and a rise in the reflectance and thereby in the lightness, as this property is dependent on the reflectance.

Basically, the dependence that hiding power shows on scattering power is also valid for the lightening power. However, the latter is more strongly system-dependent than the hiding power. Thus, evaluations of the lightening power remain limited to the test system and are not transferable to other systems.

In order to assess the lightening power, a weighed quantity (m_P) of the white pigment in question is mixed with a coloured or black paste. The amount (m_B) of a reference white pigment is then determined, which when mixed with the same coloured or black paste results in an equal lightness. If an arbitrary lightening power of 100 is assigned to the reference white pigment, the required lightening power, LP, is specified by the relationship:

$$LP = 100 \frac{m_B}{m_P} \qquad \text{at a lightness difference } \Delta Y_\infty = 0.$$

Photometric test methods for the lightening power are described in the standards DIN EN ISO 787-17 and DIN 55 982. The former specifies a fixed test-PVC of 17 %, whereas the DIN standard considers any PVC value, each one requiring a new determination. A method based on a single grey mixture was proposed by *Völz* [43].

2.3.3 Tinting strength and depth of shade

Absorption and tinting strength

The partial or complete conversion of light energy into heat by pigment particles is known as **absorption.** It may occur over the entire visible spectrum (in the case of black pigments) or selectively at particular absorption bands in this range (with colour pigments).

The ability of a pigment to absorb light depends primarily upon its absorption coefficient, and this in turn depends on chemical constitution and wavelength. In comparison, the particle size is a secondary factor. According to the theory of Mie – confirmed to be valid for pigments by *Maikowski* [44] – there is an optimum particle size for absorption, which in the case of inorganic pigments lies farther from the optimum particle size for scattering than in the case of organic pigments.

Due to the limited ability of light to penetrate pigment crystals, the core of large particles remains ineffective for absorption. Additional pigment volume accessible to the incident light is obtained by further comminution. Thus, absorption, and hence tinting strength, increase with the degree of fineness. However, if comminution proceeds beyond the optimum size, the particles allow the light to pass through and no further increase in tinting strength can occur.

The **tinting strength** of a coloured pigment is its ability to absorb the incident light and confer colour to the medium in which it is embedded (DIN EN ISO 787-24). The absorption of the medium itself is generally extremely small and can thus be neglected. Like the lightening power, the tinting strength is an indication of the yield of the colouring material. The lower the concentration of coloured pigment required to obtain a determined sensation, the higher its tinting strength.

Depth of shade

A strict distinction has to be made between tinting strength and **depth of shade**. This term refers to the intensity of a colour sensation. It increases with the saturation, but decreases with rising lightness. Whereas tinting strength is a property of the colouring material, depth of shade corresponds to the colour sensation created by a coloration, for example by a coating layer or a print.

The colorations of the coatings shown in Figure 2.16a (on page 216) were prepared with several inorganic and organic pigments [45]. They contain equal pigment concentrations and equal ratios of coloured to white pigment. Nevertheless, their depths of shade are quite different because of the fairly distinct tinting strengths of the coloured pigments. Figure 2.16b on page 216 shows the same coatings adjusted to an equal depth of shade. The corresponding ratios of coloured to white pigment provide an indication as to the tinting strength of the pigments employed.

Assessment of tinting strength through depth of shade

A quantitative determination of the tinting strength of a pigment may be performed either as an absolute value or relative to another pigment used as reference. Although visual determination of the absolute tinting strength is only possible as a rough approximation, an experienced colourist can visually determine the relative tinting strength with acceptable accuracy.

Basically, the tinting strength is assessed indirectly through the depth of shade. Reductions of the pigment under test and the reference pigment are prepared for this purpose in such a way that both reductions, when compared visually or preferably colorimetrically, can be considered to be identical. The resulting ratio of parts by mass of reference pigment to parts by mass of test pigment is a measure for the tinting strength. As a rule, the result is expressed as the parts by mass of test pigment that yield the same depth of shade as 100 parts by mass of reference pigment. Statement of the tinting strength as a percentage of the reference pigment is also possible. Thus if, for example, 90 parts by mass of the test pigment correspond to 100 parts by mass of the reference pigment, the test pigment is evaluated as 10 % stronger. A method for the visual comparison of relative tinting strength is described in DIN EN ISO 787-16.

Several colorimetric methods for the evaluation of absolute and relative tinting strength are based on the Kubelka-Munk theory. Use is made here of the fact that in white/coloured reductions, in which as a rule the white pigment is predominant, scattering by the coloured pigment is negligible in comparison with scattering by the white pigment. This justifies the assumption that absorption is governed solely by the coloured pigment, whereas the white pigment alone is responsible for scattering. Therefore, when comparing opaque layers of reductions of a test and a reference pigment at the same white/coloured ratio and prepared under otherwise identical conditions (e.g. dispersion degree), differences in the reflectance at the absorption maximum can be ascribed to the differing absorption coefficients and, consequently, to the difference in tinting strength. When using coloured pigments of high-scattering power in weak reductions, the value of the scattering coefficient of the coloured pigment may not necessarily be negligible. A possible correction of the Kubelka-Munk function for such cases is discussed by *Völz* [41].

The relative tinting strength is stated either as the ratio of the absolute tinting strength (K/S-value) of the test pigment to that of the reference pigment or by the mass of the test pigment that yields the same depth of shade as a standard amount of the reference pigment. A method founded upon this principle is described in DIN EN ISO 787-24.

The Kubelka-Munk theory is generally not applied to highly transparent systems. The tinting strength of transparent pigments, for example for four-colour printing, is assessed according to a different principle, on which the standard ISO 2846-1 is based. This standard specifies the absolute tinting strength as the pigment concentration (in percentage) that is required to obtain a previously determined depth of shade. According to DIN EN ISO 787-24, the relative tinting strength in comparison with a reference pigment is then calculated from the concentrations in per cent of sample (P) and reference (B) using the relationship:

$$F_{rel} = 100 \frac{c_B}{c_P} \quad \text{at a difference in depth of shade of } \Delta F = 0$$

Standard depth of shade

The fastness properties of pigments and of their colorations depend, among other factors, upon the concentration of the colouring material. However, use of the same concentration as a basis for comparison is inadequate for the pigment technologist, because he is interested in the fastness of specific colorations, i.e. defined depths of shade, in which the pigments are present in varying concentrations due to their different tinting strengths. In order to obtain comparable values, it is thus more sensible to test the pigments at the same depth of shade, as is shown for example in the colorations in Figure 2.16b, page 216. This requirement resulted in the establishment of shade levels termed **standard depths of shade**. DIN 53235-1 specifies depths of shade of 1/3, 1/9 and 1/25. The fractions denote the portion of pigment at 1/1 standard depth of shade that is required to attain the corresponding depth of shade. A series of concentric colour circles, each one showing 24 visually equidistant hues with the same depth of shade, is described in the DIN colour system (DIN 6164-1 and 6164-3)in which colour sectors with the same saturation and lightness are arranged on a circle. It is a peculiarity of this arrangement that the short wave colours (violet/blue) appear adjacent to the long-wave colours (red/purple). A method for the adjustment of specimens to standard depth of shade is described in DIN 53235-2.

2.3.4 Hue (or shade)

The **hue** or **shade** is the property of a coloration that differentiates it from an achromatic colour like grey or black. While saturation characterises the *degree* of colour, hue defines the *kind* of colour.

Wavelength [nm]	400	450	500	550	600	650	700
Colour	violet	blue	blue-green	green	yellow	orange	red

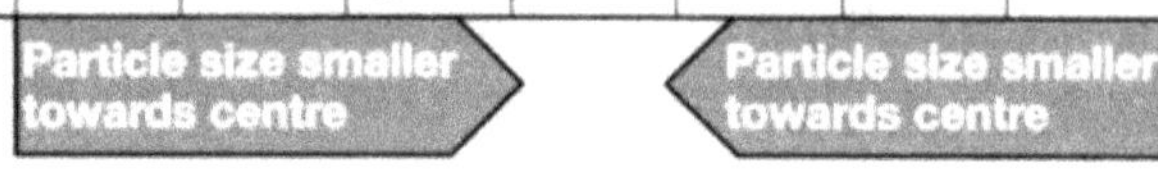

Figure 2.17: Direction of the shade change with particle size (schematic: from Kaufmann [46])

The dependence of shade on the crystal modification was mentioned in Section 1.2. A further influencing factor is the particle size of the pigment. The relationship between shade and particle size has been the subject of numerous studies, for example the articles by *Maikowski* [44], *Kaufmann* [46], *W. Carr* [47], *Hauser et al.* [48] and *Kämpf* [49] as well as a theoretical contribution by *A. Brockes* [50].

Kaufmann supports the view that a decrease in the average particle size normally causes a deviation in shade towards the centre of the visible spectrum (Figure 2.17). Hence, the shade of green pigments should only change insignificantly with diminishing particle size, a fact confirmed by the studies of *Carr.* Nevertheless, there have also been results contradicting the Kaufmann rule.

In accordance with the Mie theory, the dependence of shade on particle size arises from the dependence of absorption power on particle size and wavelength. When light strikes large pigment particles, the edges of each particle absorb radiation mainly at the absorption maximum and ensure that no more radiation at this wavelength can penetrate into the core of the particles. The radiation in the remaining wavelengths is absorbed, in contrast, only weakly at the edges and can therefore penetrate into the particle centres. Thus, in regions other than that of maximum absorption, the absorption of particles of relatively small diameter is only slightly higher than that of larger particles. This differentiated, particle size dependent absorption also gives rise to shade shiftings [51].

The brilliance and purity of a shade are influenced by scattering as well as particle size distribution. Due to the differing composition of the scattered radiation, the shade of scattering pigments appears duller than that of non-scattering pigments of the same shade. Organic pigments therefore have in general purer and more brilliant shades. For the same reason, pigments with a narrow particle size distribution yield purer shades than pigments with a wide particle size spectrum.

Apart from the factors discussed so far, the shade is also dependent upon the degree of reduction and the application medium. In printing, reductions are produced not only by mixing with a white pigment, but also by cutting down the pigment concentration in the ink or by varying the degree of effective covered area, i.e. the ratio of actual printed area to the total area. The shades of such reductions of one and the same ink carried out by different methods may, even when saturation is equal, deviate from one another. Shade deviations may also arise from different reductions of a coloured pigment. A number of red pigments show a particularly striking increase in blue tinge with an increasing degree of reduction. This phenomenon cannot be explained by the intrinsic absorption of the white pigment alone. It is rather ascribed to a scattering interaction between the red and white pigments exerting an influence on the absorption.

As to the diversity of plastics and of the natural and synthetic resins employed nowadays in practice, this has a logical consequence – the application medium may have an effect on shade. Binders and plastics can cause changes in shade not only by virtue of their own colour and refractive index, but also by the type of cross-linking, the curing conditions and their wetting properties. In many cases additives that start or accelerate the curing process may also contribute to shade variations. The influence of different curing systems on the shade stability of organic pigments in powder coatings based on epoxy resins has been verified by *Herbst* and *Hafner* [52]. They attributed the changes in shade to chemical reactions between the pigments and the relevant curing agents.

Dulog et al. studied the influence of electron beam curing on the optical properties of several classes of organic and inorganic pigments in a printing ink system [53]. They were able to show that there is no relationship between the light fastness of pigments and their stability to electron beam treatment. On the other hand, in most cases when a change in shade occurred this was found to be proportional to the applied dose of radiation. In addition, some organic pigments, e.g. dioxazine, indanthrone, hydroxy-anthraquinone pigments, exhibited distinct shade differences when the results of electron beam curing under nitrogen and under air atmospheres were compared. The authors concluded that such pigments should therefore not be used when curing is not performed in the presence of an inert gas.

If all these facts are considered, it is not surprising that efforts to introduce uniform pigmentation, i.e. one and the same pigment combination for all binder systems, were doomed to failure from the beginning

3 Rheological behaviour of pigment dispersions

3.1 Introduction

Rheology is the study of the deformation and flow behaviour of matter. This behaviour is essentially a consequence of the forces acting between the structural units of matter (atoms, ions, molecules etc.). These forces, called – for simplicity – intermolecular forces, are in their turn determined by the reciprocal distances between the structural elements, thus giving rise to an internal resistance to any change in position of the elements. Such changes are, therefore, not possible without external compulsion, i.e. without the expenditure of energy.

The intermolecular forces are particularly dependent upon the state of aggregation (it is also possible to argue the other way round), and cover an enormous range of values. The highest forces are present in the crystalline state, i.e. in a solid. This is followed by the liquid and lastly the gaseous state. In a gas at extremely low pressure, the so-called ideal gas state, the intermolecular forces are practically zero. However, intermolecular forces inside the aggregate state also depend on the kind of matter, or, in other words, they are substance-specific.

In deformation or flow processes, the rheological quantity for internal resistance is the **viscosity**. As with the intermolecular forces, viscosity also covers an enormous range. One merely has to consider the great differences in types of deformation, such as warm and cold working of metals, the flow of molten lava, the thickness of tar and the flow behaviour of gases.

Since pigment dispersions are, as a rule, systems in the liquid phase, only the rheology of these systems is of interest to the pigment technologist. The following sections are thus concerned solely with the fundamentals of rheology in the liquid phase.

3.2 Newton's law of viscosity

For a better understanding of the phenomena connected with the flow of more or less viscous substances, the following idealised arrangement may be used as a start-

Juan M. Oyarzún: Pigment Processing

ISBN 978-3-86630-665-3

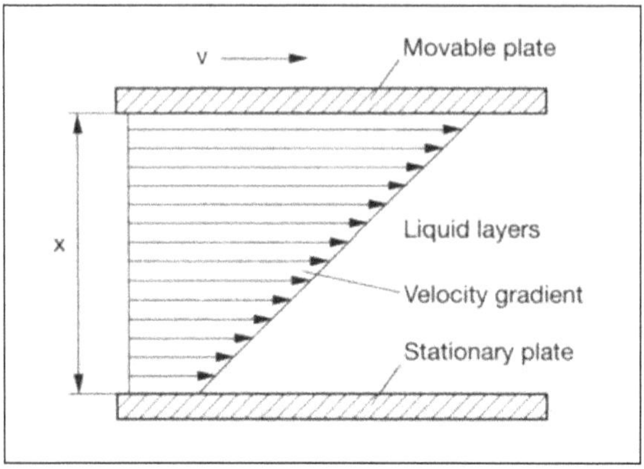

Figure 3.1: Formation of a flow-rate gradient between two parallel plates, of which one is held stationary and the other is movable. (For derivation of Newton's viscosity law)

ing point: Assuming we have a liquid enclosed between two parallel glass plates (Figure 3.1), it is obvious that a force must be applied to displace one of the plates to the side. The force required to displace the layers in the liquid is termed the **shearing force**. If the top plate is now displaced at different velocities, it will be found that the force required, *K*, and velocity, *v*, are proportional. It is evident that the force must also increase if the surface, *F*, becomes larger, and it would thus be better to express the force relative to unit area. The ratio of the shearing force to the sheared area is termed the shear stress τ.

$$\tau = \frac{K}{F}$$

Shear stress is inversely proportional to the distance x between the plates. Hence, the relationship between shear stress, distance and resulting velocity may be expressed as:

Equation 3.24:

$$\tau = \eta \frac{v}{x}$$

The factor η is a constant, and is a specific characteristic of the liquid. It is termed the **coefficient of internal friction** or **dynamic viscosity.**

As the bottom plate does not move, its velocity, and consequently the velocity of the liquid layer in contact with it, is zero. The layer adhering directly to the upper plate must move at the same velocity, v, as the plate. Between the two extremes, the liquid planes from top to bottom move at decreasing velocities because each plane is retarded by the adjacent plane directly underneath. In this process, known as laminar flow, the ratio of the difference in velocity between two adjacent layers to their distance remains constant throughout the liquid. Thus, we may express this ratio, termed shear rate or velocity gradient D, by the relationship $D = \Delta v/\Delta x$, or, as a limiting value for a layer thickness decreasing towards 0 ($\Delta x \rightarrow 0$), D = dv/dx. Hence, Equation 3.24 can be expressed as:

$$\tau = \eta \frac{dv}{dx}$$

The quantity denoted **dynamic viscosity** is, therefore, equal to the ratio of the shear stress τ to the shear rate D.

Equation 3.25:

$$\eta = \frac{\tau}{D}$$

This fundamental law of rheology was established by *Newton*. Thus, when the flow behaviour of a fluid follows this relationship, the fluid is said to be **Newtonian** and its flow properties **Newtonian** or **ideal viscous**. It can be inferred from its derivation that it is solely valid for laminar flow, i.e. without turbulence.

Expressing the shearing force in Newton (N) and the area in square meters (m^2), the shear stress is quoted in Pascals (Pa). The unit of shear rate $\Delta v/\Delta x$ in the International System of Units (SI) is accordingly m/s·m = s^{-1}, i.e. reciprocal seconds. This leads to the *Pascal·second* as the unit for dynamic viscosity (viscosity unit in SI-System: 1 Pa·s = 1 kg $m^{-1} \cdot s^{-1}$). The unit generally used previously in the CGS-System was Poise (1 Poise ~ 0.1 Pa·s). The reciprocal of the dynamic viscosity, $1/\eta$, is termed **fluidity** (symbol φ, SI-unit 1/Pa·s).

The ratio of the dynamic viscosity to the density of the liquid in question is the so-called **kinematic viscosity** ν. Its SI-unit is $m^2 \cdot s^{-1}$ (CGS-unit: 1 Stokes = 1 $cm^2 \cdot s^{-1}$). The kinematic viscosity must be taken into consideration for those flow processes in which shearing is influenced by gravitational force, as for example in efflux viscometers.

Some values for the dynamic viscosity of various liquids at 20 °C are listed in Table 3.6.

Table 3.6: Viscosity values μ in mPas at 20 °C for several fluids

Liquid	η [mPa·s] at 20 °C
Water	1.0
"Cellosolve"	1.2
Mercury	1.5
Nitrobenzene	2.2
Dibutylphthalate	19 to 22
Linseed oil	51.6
Glycerol	1,500
Tar	1,000,000

3.3 Dependence of viscosity on temperature

In general, the viscosity of a liquid decreases with increasing temperature. This is caused by weakening of the cohesion forces as a consequence of the increased mobility of the molecules at higher temperatures. A generally accepted relationship expressing the dependence of viscosity on temperature has yet to be found. However, a number of viscosity/temperature equations have been stated, although their application remains limited to specific types of liquids or temperature ranges. They are based in part on theories that relate viscosity to specific substance properties (e.g. heat of vaporisation, critical temperature). Other approaches were formulated as a result of empirical investigations, and are valid only for the range of measurements examined.

Among these equations, those of the exponential type are preferred due to theoretical considerations. For low molecular liquids the following relationship, known as the *Arrhenius equation* (also termed Andrade equation in technical literature) is in common use:

Equation 3.26:

$$\eta = A\, e^{(\Delta E/RT)}$$

where R is the molar gas constant, T the thermodynamic temperature and ΔE the molar activation energy, i.e. the energy required by the molecules in one mole to overcome the intermolecular forces and thus shift their position. The value of ΔE is usually 1/3 to 1/4 of the enthalpy of vaporisation. The constant A, which is specific for each liquid, has proved to be the limiting value of viscosity for a temperature

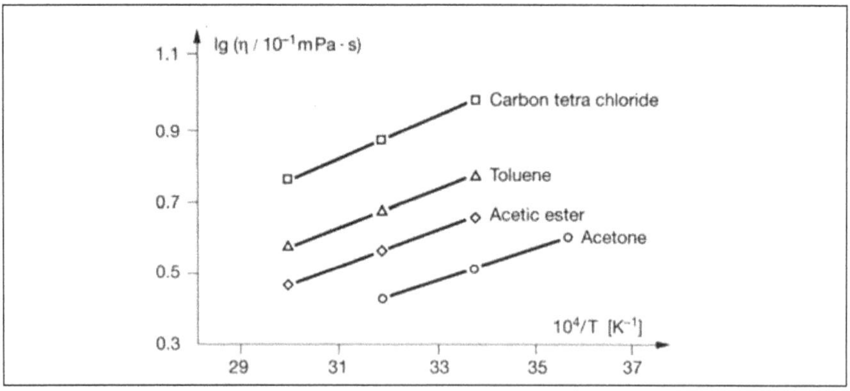

Figure 3.2: Temperature dependence of viscosity for some low-molecular liquids (plotted in accordance with the Arrhenius equation).

Source: Grinsehl Lehrbuch der Physik. B.G. Teubner Verlag, Leipzig, 1968

tending to infinity. It does not, or at least not significantly, depend on temperature. It is basically dependent on the molar mass and molar volume of the liquid.

It can easily be inferred from Equation 3.26 that, if $\ln\eta$ is plotted against 1/T, a straight line must result. This conclusion has also been confirmed experimentally. In Figure 3.2, $\ln\eta$ has been plotted for several liquids as a function of the reciprocal temperature. In most cases, straight lines are obtained with a slope corresponding to $\Delta E/R$, thus giving a measure for the activation energy.

Equation 3.26 enables the constants A and ΔE to be calculated for a given liquid with the aid of only two measurements. This may be better understood by using the following example.

Example 2

Linseed oil has a viscosity of 60.0 mPas at 10 °C and 7.1 mPas at 90 °C. Determine Equation 3.26 for linseed oil. Calculate the value for its activation energy.

Solution

Taking the viscosity values η_1 = 60.0 mPa·s and η_2 = 7.1 mPa·s and the corresponding thermodynamic temperatures T_1 = 283 K and T_2 = 363 K, the following equations may be written:

Equation 3.27:

$$\ln 60.0 = \ln A + \frac{\Delta E}{R}\frac{1}{283} \qquad (3.27a)$$

$$\ln 7.1 = \ln A + \frac{\Delta E}{R}\frac{1}{363} \qquad (3.27b)$$

Subtracting (3.27b) from (3.27a) and solving the resulting equation for ΔE/R:

Equation 3.28:

$$\frac{\Delta E}{R} = \frac{\ln 60.0 - \ln 7.1}{\frac{1}{283} - \frac{1}{363}}\,\mathrm{K} = \frac{4.094 - 1.960}{0.00353 - 0.00275}\,\mathrm{K}$$

$$\frac{\Delta E}{R} = \frac{2.134 \cdot 10^3}{0.78} = 2.74 \cdot 10^3\,\mathrm{K}$$

Constant A can be calculated by substituting the value of ΔE/R in Equation 3.27a or 3.27b and solving for ln A.

$$\ln 60.0 = \ln A + \frac{2.74 \cdot 10^3}{283}$$

$$\ln A = 4.094 - 9.682$$

$$\ln A = -5.588$$

$$A = 3.74 \cdot 10^{-3} \text{ mPa·s}$$

Equation 3.26 for linseed oil is therefore:

$$\eta = 3.74 \times 10^{-3} \exp \frac{2.74 \cdot 10^3}{T} \text{ mPa·s}$$

Calculation of the molar activation energy ΔE:

$$\frac{\Delta E}{R} = 2.74 \times 10^3 \text{ K}$$

$$\Delta E = 2.74 \times 10^3 \times 8.314 \text{ kJ/kmol} = 22.78 \text{ kJ/mol}$$

In the case of highly viscous systems, the dependence of viscosity on temperature may be better represented by equations with three or four parameters. Thus, for example, an excellent expression for the viscosity/temperature relationship is the empirically established Equation 3.29 of *Vogel.*

Equation 3.29:

$$ln\, \eta = \frac{B}{T+C} + A$$

Constants A, B and C may be determined with the aid of Equation 3.26 as shown above for constants A and ΔE. However, in this case three measurements are required. The Vogel equation allows interpolations within a large temperature range. It can be explained by the *position-shift theory*, which also accounts for the Arrhenius equation (Equation 3.26).

3.4 Flow properties of concentrated suspensions and molecular colloids

High concentration of the dispersed phase in a dispersion (e.g. pigment pastes, pigmented printing inks) or of the macromolecules in a molecular colloid gives rise to interactions between the particles or macromolecules. The flow properties of such systems can no longer be described using the relationships discussed previously. The intermolecular forces taking effect in these cases result in a series of phenomena, termed **anomalous flow properties**, which will be briefly dealt with in the following sections.

There are a large number of fluids and pastes with flow properties deviating from Newtonian. Their viscosity coefficient is not only dependent on temperature, but is also a function of shear rate and/or shearing time. These substances are termed **non-Newtonian**, and their viscosity function η_D is defined as the differential quotient $d\tau/dD$ at a given shear rate. Such **anomalous flow behaviour** is exhibited by molten plastics, liquid crystals and colloidal solutions and suspensions at higher concentrations. In such cases, one single determination of the viscosity coefficient at a given shear stress is not enough to characterise the flow properties of the system. The dependence of viscosity on rate of shear must be determined over the whole range of shear rates which are of practical interest. For this purpose, the shear-stress values τ measured are plotted against the corresponding rates of shear D (the reverse is also possible). A **rheogram** or **flow curve** is obtained. Plotting the viscosity function in dependence of shear stress or shear rate, termed a viscosity curve, is also normal practice.

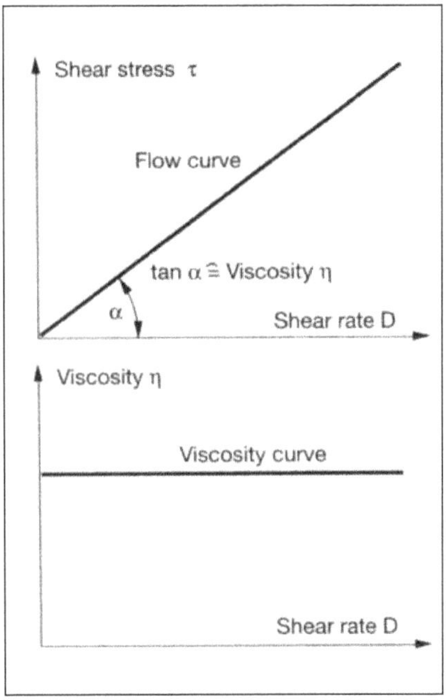

Figure 3.3: Flow curve (above) and viscosity curve (below) of a Newtonian liquid (schematic). The viscosity coefficient is a constant, i.e. not dependent on the velocity gradient and corresponds to tan α when the unit segments on both axes are equal.

3.4.1 Newtonian flow

The viscosity of Newtonian or ideal fluids, as defined in Section 3.2, is a material constant that depends only upon temperature and (generally to a negligible extent) on pressure. One single measurement is therefore sufficient for its characterisation at a given temperature. The flow curve as well as the viscosity curve are straight lines (Figure 3.3) in this case. The slope of the straight line in the τ/D-diagram, i.e. tan α, corresponds to the viscosity coefficient.

Liquids with a simple molecular structure, e.g. water, organic solvents,

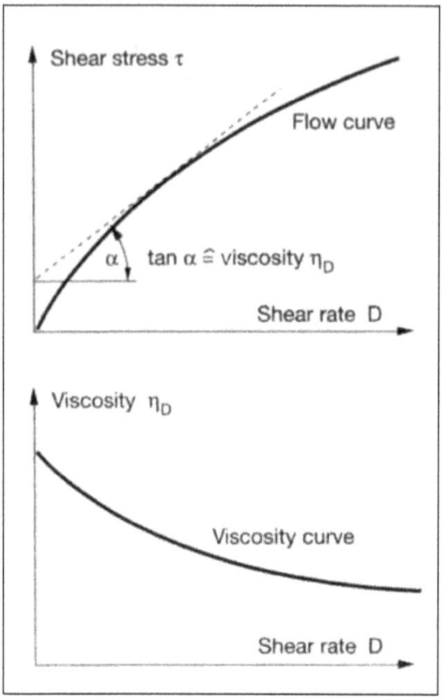

Figure 3.4: Flow behaviour of a pseudoplastic substance. Flow curve (above); viscosity curve (below). The viscosity decreases with increasing velocity gradient

linseed oil, glycerol etc., are Newtonian. Polymer solutions at low concentrations, varnishes and even a large number of pigmented systems, such as paints and printing inks diluted to application conditions, also follow Newton's law of viscosity in practice. Molten plastics at extremely low shear rates similarly behave as ideal fluids.

3.4.2 Structural viscosity (pseudoplasticity)

Some substances grow thinner when submitted to shearing forces, i.e. their viscosity decreases with increasing shear stress. In this case, the relationship between shear stress and shear rate is not linear, but obeys rather a power law. Such substances are called **pseudoplastic**. The graphs in Figure 3.4 show their typical flow and viscosity curves.

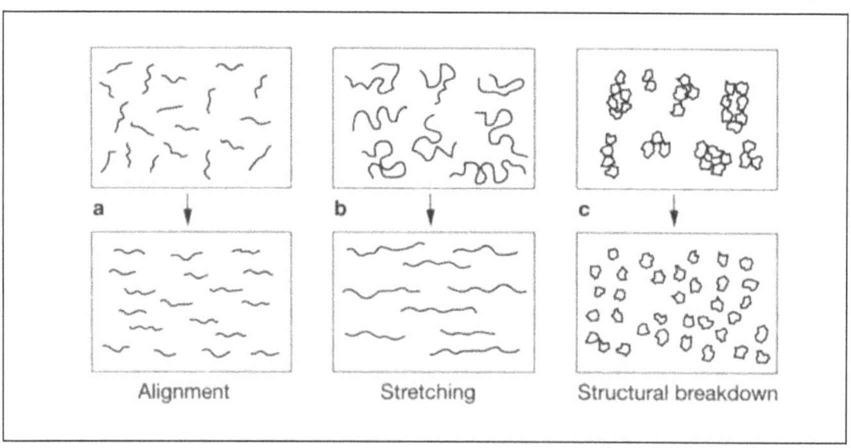

Figure 3.5: Structural changes in polymer solutions or dispersions giving rise to pseudoplastic flow behaviour: (a) alignment of thread-shaped molecules; (b) stretching of coiled molecules; (c) breakdown of micelles and flocculates

Structural viscosity is a characteristic of systems with particle interaction, i.e. systems containing coiled molecules, micelles or flocculates. The viscosity decrease is caused by structural changes brought about by shearing. Examples are the alignment of thread-shaped molecules in the direction of flow (Figure 3.5a), the stretching of coiled molecules (Figure 3.5b) and the breakdown of micelles and flocculates (Figure 3.5c).

A further cause of viscosity decrease under shear can be the disengagement of sterically attached or bound solvent molecules by stretching or particle deformation. In both cases, part of the solvent becomes free to join the flow.

Flow curves of pseudoplastic materials may differ considerably from one substance to another. Even the same substance may, at varying concentrations or in different solvents, give rise to flow curves with very unequal courses. Empirical relationships for characterising pseudoplastic behaviour are thus not always satisfactory and have only a narrow range of validity. The Ostwald equation is an expression often used that at least approximately describes the flow curve:

Equation 3.30:

$$\tau = k\,D^{n} \qquad \text{with } 0 < n \leq 1$$

The parameters k and n are system-dependent and may be determined empirically. The viscosity function at an arbitrary shear rate D is equal to the first derivative and corresponds to the value of tan α, provided that the unit segments on both coordinates are equal (Figure 3.4 above). In case of $D = f(\tau)$-type graphs, the viscosity function at an arbitrary shear rate is, of course, given by cot α.

$$\eta_D = \frac{d\tau}{dD} = k\,n\,D^{n-1} \triangleq \tan\alpha$$

The viscosity curve is formally represented by the differential curve of the function $\tau = f(D)$, i.e. with the graph of the viscosity function η_D versus the shear rate D (or shear stress τ). If the pertinent equation is not available, this can be performed by graphical differentiation, i.e. by determining the slope (tan α) of the flow curve at some particular D-values and plotting tan α against D.

Example 3

The following values for shear rate and shear stress were obtained during viscometric testing of a pseudoplastic paint at 23 °C:

D [s^{-1}]:	5	10	15	20	25	30	35	40	45	50
τ [Pa]:	15.6	24.3	30.2	34.8	38.8	41.6	44.2	46.4	48.2	49.6

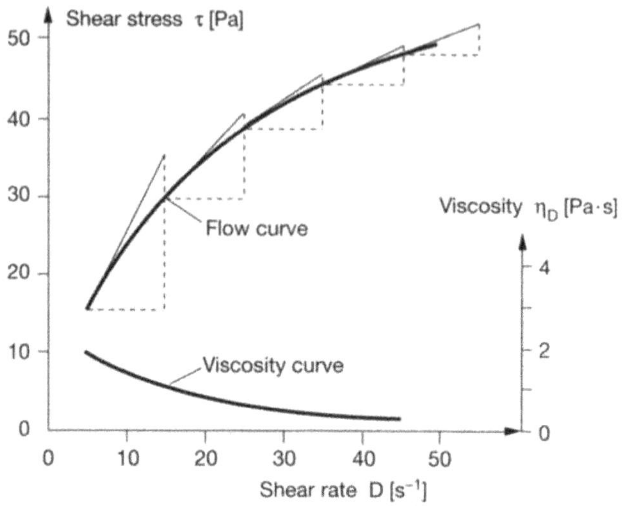

Figure *3.6: Example 3: flow curve (above) and corresponding viscosity curve determined by graphical differentiation (below)*

Solution

In Figure 3.6, the flow curve (above) and the corresponding viscosity curve assessed by graphical differentiation (below) can be seen. This method of differentiation is described in various textbooks on the relevant mathematics [54, 55].

In the pertinent English literature, the viscosity at an arbitrary shear rate η_D is termed **apparent viscosity**. Although apparent in this context does not mean that the measured viscosity is imaginary but rather that the viscosity corresponds to only one shear stress value, for the sake of clarity it would be advisable to replace this designation by the new, less ambiguous expression **representative viscosity**, as specified in DIN 1342-1 or in DIN EN ISO 3219.

Structural viscosity is exhibited by suspensions, concentrated polymer solutions, latex paints, molten polymers, highly pigmented paints and printing inks, lubricating oils with additives, adhesives and pigment pastes.

3.4.3 Dilatancy

In some substances the viscosity increases with increasing shear rate. This flow behaviour, known as **dilatancy** and appearing opposite to structural viscosity, is caused by an interaction between molecule coils due to partial valency forces impairing the mobility of the chains and hence the flow. However, at very high concentrations of the dispersed phase, its total surface area binds a relatively large amount of the continuous phase, thereby also hindering the particles' mobility. Submitted to low shearing conditions, the suspension behaves as a fluid, but at high shear ideal flow is no longer possible due to the mutual hindrance between particles. At conditions of extreme strain, the practically infinite increase in viscosity may even cause a rupture in the system.

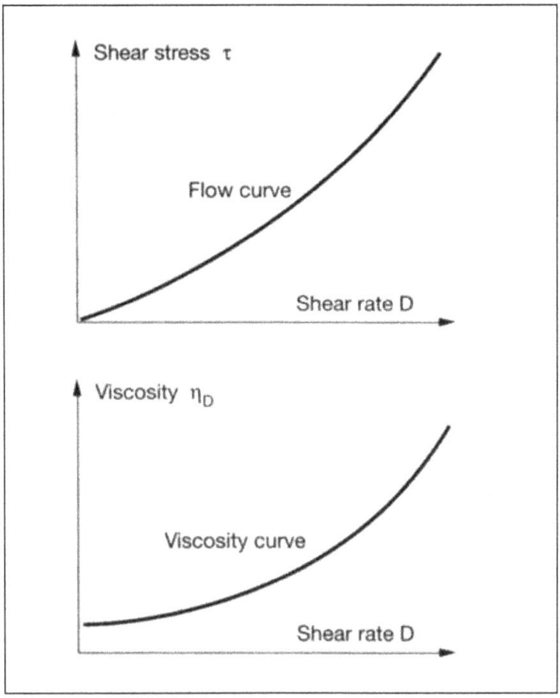

Figure 3.7: Flow behaviour of a dilatant substance. Flow curve (above); viscosity curve (below). The viscosity increases with increasing shear rate

The process taking place during shearing can be reversible, i.e. after consolidation of the system, flow may start again. The τ/D rheogram and viscosity curve of dilatant substances show a line bending towards the vertical co-ordinate (Figure 3.7).

Dilatancy can also be approximately described by the Ostwald equation (Equation 3.30), n being in this case higher than 1. However, this relationship is only valid for a narrow range.

Dilatant flow occurs more seldom than pseudoplasticity. Highly concentrated suspensions, silicones, plastisols, starch in water, sand slurries and occasionally also pigment pastes exhibit this flow anomaly.

3.4.4 Plasticity

Many substances do not flow immediately when subjected to shearing forces, but only after a minimum shear stress, the so-called **yield value** or **yield point**,

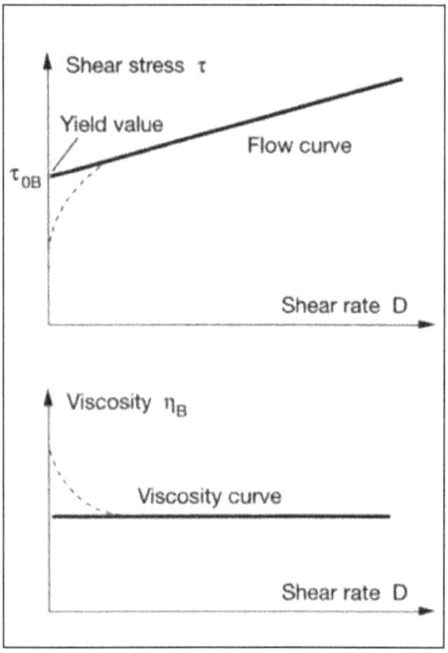

Figure 3.8: Rheograms of a linear-plastic substance: flow curve (above); viscosity curve (below). Once the yield value τ_{OB}, has been exceeded, rate of shear is proportional to shear stress, i.e. viscosity coefficient is a constant

has been attained. Such substances, for the most part dispersions, are qualified as **plastic**. The yield value arises from intermolecular forces, such as hydrogen bonds, van der Waals and electrostatic forces, which induce the dispersed phase to build up a structure. In many cases such structures encase and immobilise large amounts of the liquid phase, and they behave like elastic bodies under low shearing forces, i.e. below the yield value. The often gel-like structure will only break down and allow the system to flow when the shearing forces are high enough to overcome the forces holding the structure together. The yield value is thus a measure for the gel stability of the system.

Plastic substances may be classified into three groups: Bingham or linear-plastic, non-linear plastic (Casson) and plastic-dilatant systems.

3.4.4.1 Bingham or linear-plastic flow

A number of liquids show Newtonian flow behaviour after the shear forces have overcome the yield value, i.e. when flow sets in, shear stress and shear rate are proportional to each other. The flow curve of these systems is therefore, as in the case of Newtonian systems, a straight line. The only difference is that its origin lies at a point on the τ-axis above 0. The same applies to the viscosity curves (Figure 3.8).

These substances are often referred to as **Bingham bodies,** and their flow behaviour as **Bingham viscosity**, which is characterised by the equation:

Equation 3.31:

$$\eta_B = \frac{\tau - \tau_{OB}}{D}$$

where τ_{OB} is the **Bingham yield value**.

The yield value of these systems is not always as clear-cut as the intersection of the straight line with the τ-axis in Figure 3.8 (above) would suggest. In many cases, breakdown of the structure does not take place abruptly, but stepwise, so that a certain transition stage between elastic body behaviour and perfect flow may be observed. This transition stage is represented by the dashed line in Figure 3.8.

Typical Bingham bodies are various ceramics materials, butter, oil paints, tooth-paste and lipsticks. Some types of emulsion paint also show Bingham flow properties.

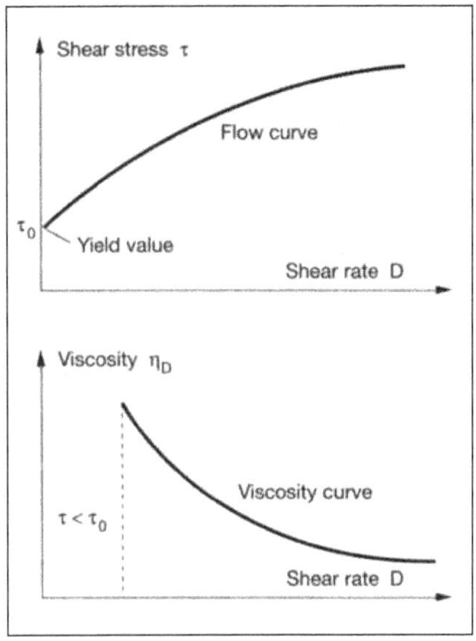

Figure 3.9: Flow behaviour of a non-linear plastic substance: flow curve (above); viscosity curve (below). After overcoming the yield value, the substance behaves purely pseudoplastically.

3.4.4.2 Non-linear plasticity (Casson viscosity)

In addition to this anomaly, a great many pseudoplastic systems exhibit a gel stability that is high enough to result in a yield value. Typical rheograms for this flow behaviour are presented in Figure 3.9.

According to *Casson*, this type of flow can be characterised by the relation:

Equation 3.32:

$$\eta_\infty{}^n = \frac{\tau^n - \tau^n}{D^n}$$

where η_∞ = final viscosity. This relationship, which was initially determined empirically, may also be explained theoretically [56].

The **final viscosity** is the viscosity value extrapolated to extremely high shear rate. In this region, all flow anomalies resulting from particle interaction or other causes are practically eliminated. The exponent n governs the degree of curvature. As a general rule, systems with a high pigment concentration (e.g. mill bases, pigment concentrates, printing inks) may be characterised using a n-value of 0.5. In this case, the Casson relationship (Equation 3.32) for these systems becomes:

Equation 3.33:

$$\sqrt{\eta_C} = \frac{\sqrt{\tau}-\sqrt{\tau_{0C}}}{\sqrt{D}}$$

or, after transposition:

$$\sqrt{\tau} = \sqrt{\eta_C}\sqrt{D}+\sqrt{\tau_{0C}}$$

Hence it follows that a plot with $\sqrt{\tau}$ instead of τ as the ordinate and $\sqrt{D}$ instead of D as the abscissa should yield a straight line (Figure 3.10). Its slope corresponds to $\sqrt{\eta_C}$, and its intercept on the $\sqrt{\tau}$ – axis gives $\sqrt{\tau_{0C}}$, thereby allowing determination of the Casson yield value τ_{0C}. The viscosity coefficient η_C is termed **Casson viscosity**.

It can occasionally be observed that the lower part of the flow curve in the Casson plot is not a straight line, but a curve running asymptotically to the ordinate axis, i.e. there is no actual yield value (Figure 3.10 below). In these cases, the τ_{0C}-value obtained by extrapolation is termed the **fictitious yield value**. It is nevertheless important for the practice, as the pigment technologist can be sure that the system will exhibit flow above this value.

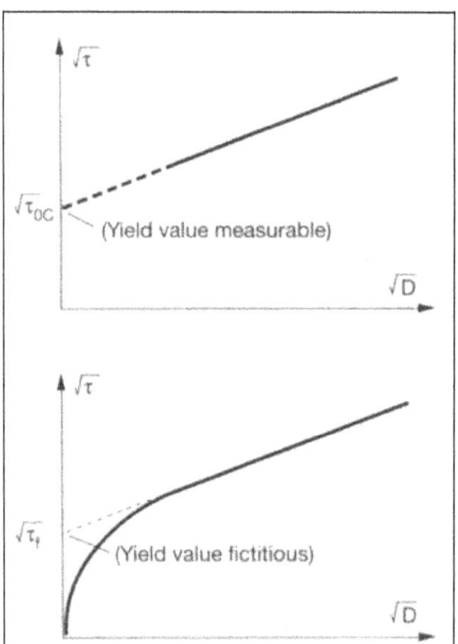

Figure 3.10: Rheological characterisation of a non-linear plastic substance according to Casson, with measurable yield value τ_{0c} (above) and with fictitious yield value τ_f (below)

Another model for describing the flow behaviour of this kind of system is the *Herschel-Bulkley* relationship [57]:

Equation 3.34:

$$\tau = \tau_0 + \mathrm{k}\, D^n$$

This actually represents a modified *Ostwald* equation (Equation 3.30), in which a yield value is taken into consideration. The condition $n < 1$ also applies in this case. The plot of τ against D^n results in a straight line, whose slope corresponds to the parameter k. The intercept on the ordinate axis is τ_0.

Indirect verification of the yield value

In extremely low ranges of *D*, rheological measurements are very difficult, and a mistake can lead to the assumption that there exists a yield value, i.e. the system is erroneously regarded as a Casson system. For this reason, *Hadjistamov* recommends plotting *t* against *D* on double-logarithm paper [58]. As a rule, plastic systems show in this kind of graph one or several straight lines, the slopes of which increase with the shear rate. If low enough shear rates are achievable, extrapolation of the straight line with the lowest slope towards the shear stress axis allows to verify whether or not the material exhibits a yield value.

Example 4

On studying the rheological behaviour of a mill base containing 18 % by weight of perylene red (C.I. Pigment Red 179) at 23 °C using a rotational viscometer, the following values for shear stress and shear rate were obtained:

D [s^{-1}]:	2.13	4.11	7.91	15. 24	23.59	36.53	56.55
τ [Pa]:	67.0	83.9	106.9	134.8	155.0	177.5	200.5

Is this mill base a pseudoplastic or a non-linear plastic system?

Solution

In Figure 3.11 the flow curve is plotted linearly on the left and double logarithmically on the right. In the latter case, it may be concluded from the intercept of

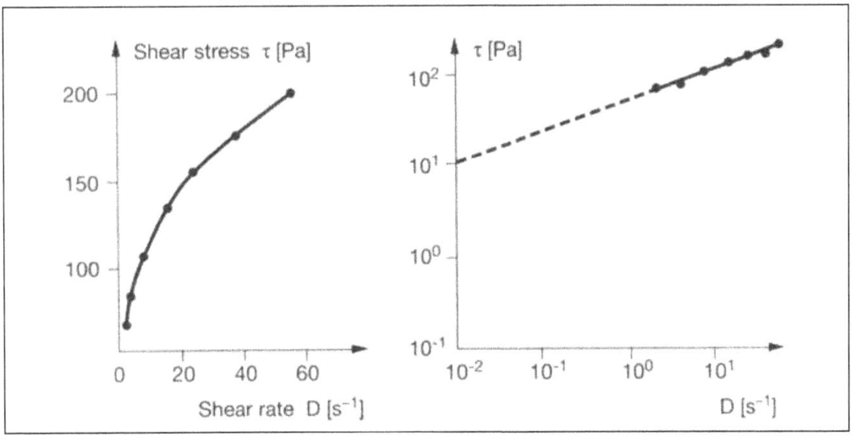

Figure 3.11: Example *4: Flow curve in linear (left) and double logarithmic plotting (right)*

the straight line with the shear stress axis that there is, even at an ultralow shear rate, still a considerable shear stress. The system should therefore be regarded as non-linear plastic.

Most of the non-linear plastic substances can be adequately characterised by **yield value** and **final viscosity**. In the case of pigment dispersions, the yield value is inversely proportional to the particle size, i.e. it increases with progressive dispersion. The yield value is also dependent upon the interaction between pigment particles and dispersion medium. The final viscosity is governed by pigment volume concentration and the viscosity of the medium [48]. See also Section 3.5.

Fillers, emulsion paints, plastisols and a large number of pigmented systems, e.g. polyvinyl chloride spread-coating pastes, paints and printing inks for letter press and offset printing are non-linear plastic materials.

3.4.4.3 Plastic-dilatant flow

The opposite of non-linear plastic flow, i.e. an increase in viscosity with increasing shear stress after the yield value has been exceeded, is named plastic-dilatant flow. This anomaly has been found in some gels. It is assumed that in these systems the absence of electric charges on the particles of the disperse phase enables the formation of a yield value. Aside from this fact, the reasons for the viscosity increase are the same as for dilatancy. The causes of dilatant and plastic-dilatant flow have not been completely elucidated as yet.

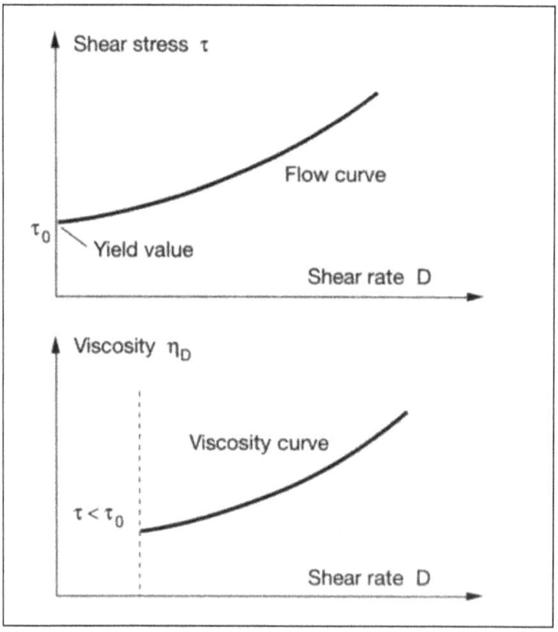

Figure 3.12: Rheograms of a plastic-dilatant substance: flow curve (above); viscosity curve (below). After overcoming the yield value, the viscosity increases with increasing shear stress.

The types of flow and viscosity rheograms of these systems are shown schematically in Figure 3.12. Reasonable approximation values are obtained in this case by applying the Herschel-Bulkley relationship (Equation 3.34) with the condition that $n > 1$. Plastic-dilatant flow systems are seldom encountered.

3.4.5 Shear-time dependent flow

In the systems with anomalous flow examined up to now, the viscosity varied only with temperature and shear stress or rate. However, systems are often encountered whose viscosity at a constant rate of shear changes with the duration of the strain. Depending on the kind of change, i.e. decrease or increase in viscosity, the system is called **thixotropic** or **rheopex** respectively.

3.4.5.1 Thixotropy

Functional basis and characterisation

Thixotropic materials submitted to a fixed mechanical strain show a decrease in viscosity with time. When the strain is removed, the system regains its original viscosity after resting for some time. This kind of flow is attributed to the action of intermolecular forces at the surface of the disperse phase. Under their influence and as long as the fluid is allowed to remain at rest, the particles coalesce and form a three-dimensional structure. If the system is now subject to constant shearing, this structure is gradually destroyed (breakdown of gel). Once at rest again, the structure starts reforming once more. Consequently, we are dealing here with a reversible gel/sol-transformation. Schemes describing thixotropy are shown in Figure 3.13.

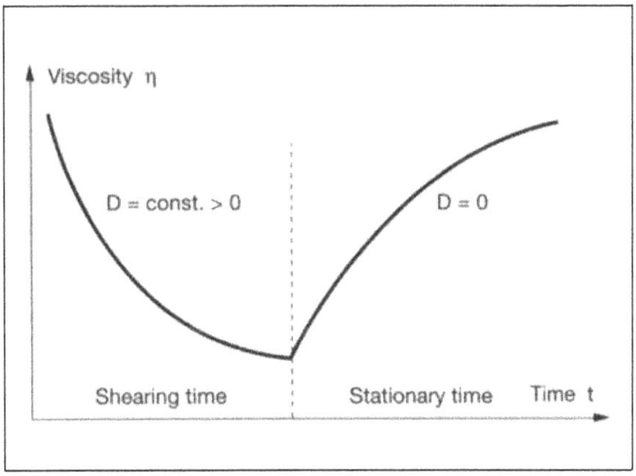

Figure 3.13: Change in the viscosity of a thixotropic system with time under constant shear (left) and at rest (right)

Since this flow anomaly is time-dependent, measurements solely of shearing stress and rate of shear are insufficient for its characterisation. The best approach to gauge the thixotropic behaviour of a system is to determine its **thixotropic loop**. For this purpose, the substance is first examined at gradually increasing shear rates. A series of related values for D and τ are thus obtained and allow the **gel-curve** to be plotted. The highest shear rate is allowed to act upon the system for a certain period in order to secure the collapse of the gel structure. The measurement is then repeated, but this time at gradually decreasing shear rates. The new pairs of τ- and D-values produce the **sol-curve**. It will be noticed that the same D-values now result in lower τ-values, or, in other words, the sol-curve and the gel-curve do not coincide. The result is illustrated with $D = f(\tau)$ in Figure 3.14.

The area delimited by the hysteresis loop, i.e. by the sol- and gel-curves, and by the dashed line in Figure 3.14 can be regarded as a measure of the degree of thixotropy of the system. However, the result and hence the delimited area depend not only on the examined system, but also on its **previous history** and on the **test conditions**, e.g. duration and extent of the shear strain, so that it is not easy to obtain accurate and reproducible values. Moreover, the fact that pigmented systems often show several flow anomalies at the same time does not make the study of their rheological behaviour any easier. A method for determining the hysteresis loops of water-thinnable paints, the **rotation test with ramp**, was proposed by *Schmidbauer* [59]. It would be worthwhile to consider its application to other systems using suitable shearing ranges and times.

Practical consequences

The important factor "time" is a good means to differentiate thixotropy from pseudoplastic flow. Whereas for the latter the original state is reached relatively soon after the strain is removed, in the case of thixotropy this requires a certain regeneration time which may possibly be quite long.

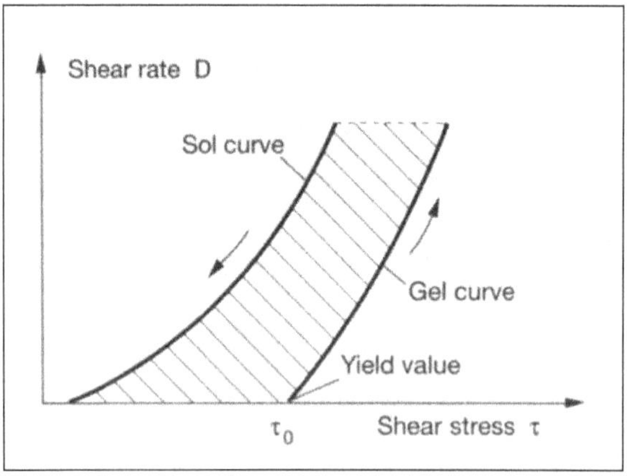

Figure 3.14: Gel and sol curves of a non-linear plastic substance with thixotropic properties.

In many cases thixotropy may be considered as a desirable attribute, because important application properties of a system, e.g. levelling and storage stability, can be favourably influenced by regulating its thixotropy. This is performed by dissolving or dispersing special additives in the system, the so-called **thickening agents**, which influence its thixotropy either by swelling or by building a gel-like structure. More details on this topic are given in Section 3.5.3.

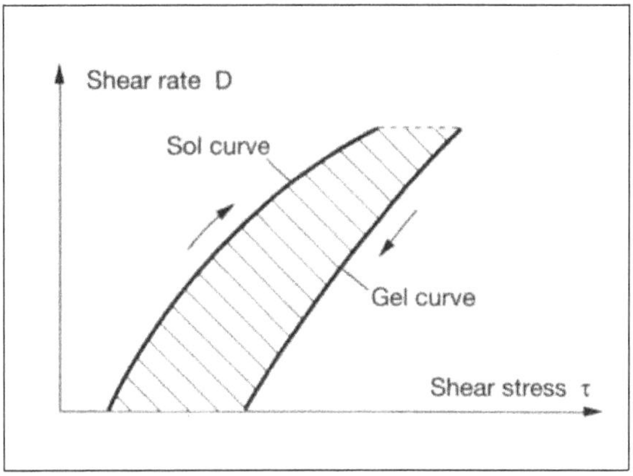

Figure 3.15: Hysteresis loop of a rheopectic substance

Thixotropy is very often encountered in suspensions of polar materials in non-polar fluids. Because the dispersed particles are insufficiently wetted, they show a strong tendency to re-agglomerate. Some high-polymers, e.g. cellulose glycolate, are also able to build gel-like thixotropic structures due to molecular interaction.

Thixotropy occurs in pastes (in tubes), latex paints, organosols, alkyd resin paints, high viscosity oil inks and flushed pigments.

3.4.5.2 Rheopexy

The time-dependent thickening of a system under a constant mechanical stress, i.e. the inverse phenomenon to thixotropy, is called **rheopexy**. This anomaly, also a reversible process, can only be observed at low rates of shear, as gel formation cannot take place at very high stresses. A possible explanation for this behaviour is that the system is in a state of non-equilibrium, and can attain equilibrium by stirring or shaking. In the rheogram (*D* as a function of τ), the hysteresis loop of a rheopectic substance runs clockwise, the reverse of thixotropy, i.e. the up curve (sol-curve) is superimposed on the down curve (gel-curve) (Figure 3.15).

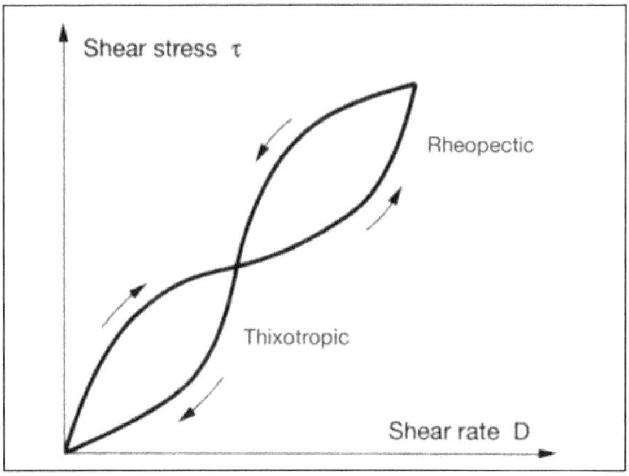

Figure 3.16: Hysteresis loop of a substance with thixotropic and rheopectic flow behaviour

Rheopexy is seldom encountered in pigmented systems. In contrast to thixotropy, only few examples of rheopectic substances can be quoted. This is probably the reason why rheopexy has not been thoroughly investigated as yet. Vanadium oxide suspensions and soap soles exhibit rheopectic flow.

Materials have also been found that can exhibit thixotropy and rheopexy, depending on the shear stress range. In such cases, the hysteresis loop shows an inversion (Figure 3.16).

3.4.6 Viscoelasticity

3.4.6.1 Effect principles

In two or multi-component systems, e.g. suspensions and polymer solutions, it may occur that one phase behaves like a fluid, whereas in the other phase more elastic properties come into play, i.e. under a shearing strain its reaction corresponds to that of a solid. In such systems, termed **viscoelastic**, both viscous (irreversible) and elastic behaviour can be encountered simultaneously. However, homogeneous systems, e.g. molten polymers, may also show viscoelastic flow properties when their molecular interaction enables the formation of both time- and shear-dependent association states. Partly crystalline polymers are a good example, as their molecules adhere more strongly in the crystalline regions than in the amorphous ones. As a consequence, the mobility of the particles in this type of plastic material lies between that of crystals and that of liquids, resulting in viscoelastic flow.

The deformation of a viscoelastic material subjected to a mechanical strain depends on the rate of shear. Under very low shear rates it behaves predominantly as a fluid, and when the shear rate is extremely high, mainly as a solid, i.e. elastic. Hence it follows that it is possible to consider the energy of deformation in two separate parts: The first, viscous part is dissipated as frictional heat. The second part causes the elastic deformation and is reversibly stored in the deformed condition. It is therefore quite obvious that an equation for mathematically expressing viscoelastic deformation can be derived by combining the relevant mechanical laws for elastic and viscous processes.

Quantitative relationships

The elastic part obeys Hooke's law, according to which the shearing, i.e. the deformation, is proportional to the stress applied.

$$\tau = G \cdot \varepsilon_{el}$$

where τ = shear stress

G = shear modulus, the proportionality constant,
ε_{el} = elastic deformation, i.e. shearing.

In an element of time dt, the increase in elastic deformation is:

$$\frac{d\varepsilon_{el}}{dt} = \frac{1}{G}\frac{d\tau}{dt}$$

The viscous part follows Newton's law of viscosity for ideal viscous systems (Equation 3.25), which is quoted here again to simplify matters:

$$\tau = \eta D \qquad \text{or} \qquad D = \frac{\tau}{\eta}$$

The factor 'time' is already taken into account in Newton's relationship. The rate of shear coincides with the velocity of deformation. Thus the total deformation in a time element is equal to the sum of both components of the deformation:

Equation 3.35:

$$\frac{d\varepsilon}{dt} = \frac{1}{G}\frac{d\tau}{dt} + \frac{\tau}{\eta}$$

According to this differential equation, which was deduced by *Maxwell*, the velocity of deformation of a viscoelastic substance is proportional to the variation of shear stress with time and to the shear stress itself. The term τ/η is the **creep**

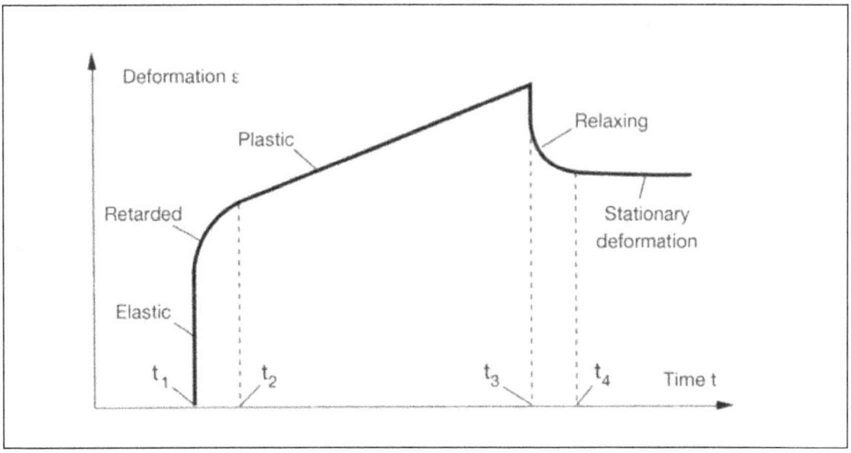

Figure 3.17: Deformation of a viscoelastic material under constant shear strain as a function of time (see text for further explanations)

velocity. Viscoelastic substances are also called **Maxwell bodies**. Therefore, Equation 3.35 is the rheological equation for the status of a Maxwell body.

The scheme in Figure 3.17 illustrates the deformation progress of a viscoelastic material subjected to a shearing strain as a function of time.

The shearing strain starts at the instant t_1 and mainly gives rise to an elastic deformation that is independent of time. However, it proves not to be absolutely independent of time when a longer period of strain is considered. Due to the **elastic after-effect**, part of the deformation occurs after a certain period of delay. This is illustrated between the times t_1 and t_2. In this interval, a plastic, irreversible part of the deformation occurs, termed **creep**, and the material behaves like a liquid or, more generally, as a plastic substance. The stress may be removed at time t_3. The elastic deformation disappears relatively quickly (relaxation) and the remainder is the irreversible, stationery deformation. The interval t_4 - t_3 is termed the relaxation time.

3.4.6.2 Effect of molecular structure

The elastic properties of rubber-like polymers can be fundamentally ascribed to changes in their **entropic state**. When at rest, the polymer chains of melts or in plastic solutions are intertwined with one another completely at random. This arises from the fact that they have a tendency to assume the most probable configurations from a purely statistical point of view, or from interactions either mutual or with the solvent molecules. When such materials of high molecular weight are submitted to a shearing strain, this results in uncoiling, stretching and alignment of the polymer chains. Since the coiled, unstretched configuration represents a state of disorder, shearing the polymer chains results in a more ordered arrangement and, as is commonly known, in a decrease in entropy. The polymer chains thus revert to their original condition, i.e. that of higher entropy, as soon as the shearing stress ceases [60].

A typical manifestation occurring in viscoelastic materials is the so-called **Weissenberg effect**, i.e. the more or less paste-like material moves up the shaft of the stirrer under strong shear. Another example is the swelling of polymer melts when flowing out of a nozzle. The Weissenberg effect is due to normal stresses arising in a visco-elastic system together with shear stresses when the material is intensively stirred. The swelling behaviour is the result of the entropy changes mentioned above. In detail, the polymer chains, forced to stretch and align by the flow, will find sufficient space to revert to their random condition again after leaving the nozzle, implying a volume increase of the extruded strand. This and similar cases are also referred to as **entropic elasticity**.

As a rule, pigmented systems also show viscoelastic behaviour. This is strongly dependent on the polarity of the pigments and their concentration. Viscoelastic properties can be particularly noticed during the dispersion process or during the application of printing inks by high-speed rollers (see Section 3.5).

3.4.7 Flow behaviour of water-thinnable binder systems

Although water-thinnable polymer solutions and dispersions behave in many cases similarly to the systems discussed above, their flow properties exhibit certain peculiarities and they occupy a special position from the rheological point of view.

Water-soluble polymers are polyelectrolytes, i.e. they consist of macromolecules possessing polar groups such as hydroxyl and ammonium groups. At comparable molecular weights and solid contents, polyelectrolyte solutions exhibit a higher viscosity than resin solutions in organic solvents. This has a detrimental effect on their application possibilities. The characteristic feature of these polymer solutions is their behaviour during thinning. After an initial decrease, their viscosity increases again due to the fact that solvation results in the formation of micelles (Figure 3.18). This anomaly can be avoided by the addition of organic solvents or by changing the chain structure.

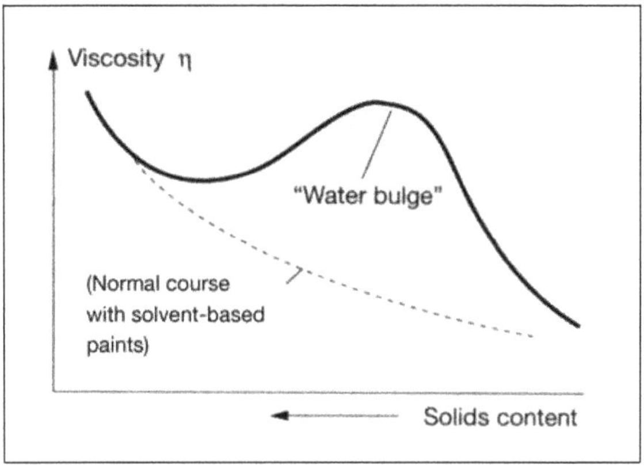

Figure 3.18: Change in the viscosity of an aqueous molecular colloid (waterborne paint) on dilution

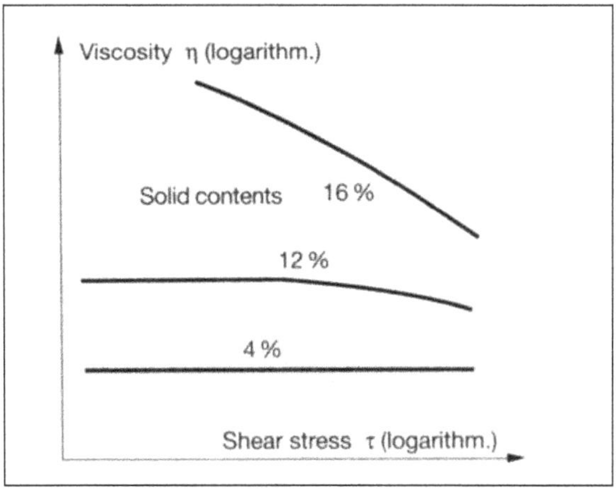

Figure 3.19: Logarithmic plot of the viscosity change with shear stress in waterborne polymer dispersions at different solid contents

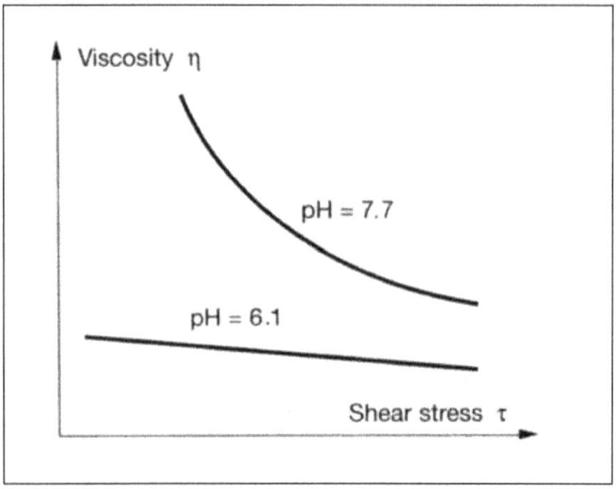

Figure 3.20: Viscosity change in waterborne polymer dispersions with shear stress at different pH-values

Water-thinnable polymer dispersions do not have these disadvantages. Since here the molecular weight has no effect on viscosity, polymers with high molecular weight can be used as the disperse phase, resulting in dispersions with high solid contents. The flow properties are influenced by concentration, pH value and the ratio of water to solvent in the dispersion medium. The dependence of viscosity upon solid contents and pH value is shown in Figure 3.19 and 3.20, using a **core-shell dispersion** as a model example.

The reason for the viscosity rise with increasing solid contents is the increased interaction between the dispersed particles. The transition from an acid to an alkaline pH range causes a solvation of the polymer layer, giving rise to pseudoplastic flow [61].

3.5 Rheological behaviour of pigmented systems in practice

3.5.1 Basic aspects

The flow properties of pigmented systems are of great importance for their production as well as for their storage and application. Pigment-free varnishes and resin solutions for paints and printing inks behave as a rule as Newtonian fluids. Conversely, high-solids and water-thinnable media show the characteristic flow properties of high polymers. Under application conditions, many systems containing low pigment concentrations exhibit virtually ideal-viscous flow. Generally speaking, however, pigmented systems deviate from ideal-viscous behaviour and have flow anomalies that frequently overlap. The application of mathematical rules to characterise their flow behaviour is thus a very difficult matter. Strictly speaking, this is only possible with Newtonian fluids. In all other cases, as often mentioned in the preceding sections, mathematical relationships can only be applied as makeshift approximations.

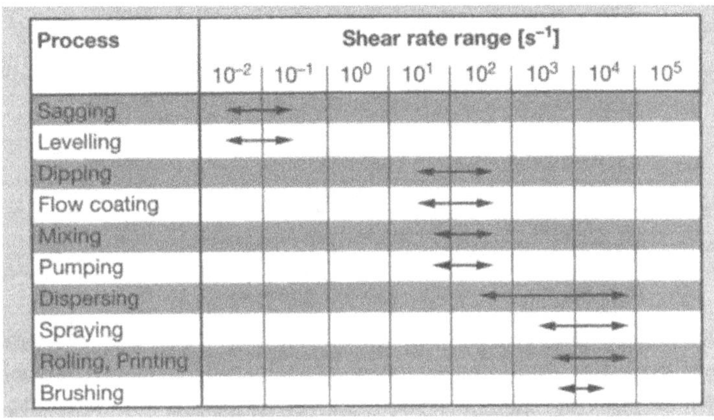

Figure 3.21: Some average shear rate ranges occurring during manufacture, storage and application of pigment dispersions

As explained previously, flow anomalies are strongly dependent on shear rate. The arrangement in Figure 3.21, compiled by *Pierce*, gives an impression of the ranges of shear rates that occur in the production, storage and application of pigmented systems [62].

The logical consequence of such a considerable span of shear stress is that the rheological variables of pigmented media during manufacture, storage and application, depending on the shearing conditions, may assume values also varying within an enormous range. If, in addition, the diversity of flow anomalies are taken into account, it is no wonder that the rheology of non-Newtonian substances was based only on practical facts and purely empirical results for a long time. However, the rapid progress in measurement and testing methods during the past decades has enabled the establishment of a scientific approach built up on firm theoretical foundations.

Practical consequences

The following section is related to the practical significance of the flow behaviour of pigment dispersions during their application and storage. The dispersion process itself and the factors influencing it (including rheology) will be the subject of a separate section in this monograph (see Section 6.1).

Pigments dispersed in printing varnishes, preparations, plasticisers or resin solutions influence the flow properties by virtue of their surface character, particle shape, specific surface area and concentration. Particle size distribution is, on the contrary, as far as the flow properties are concerned, of no importance [63].

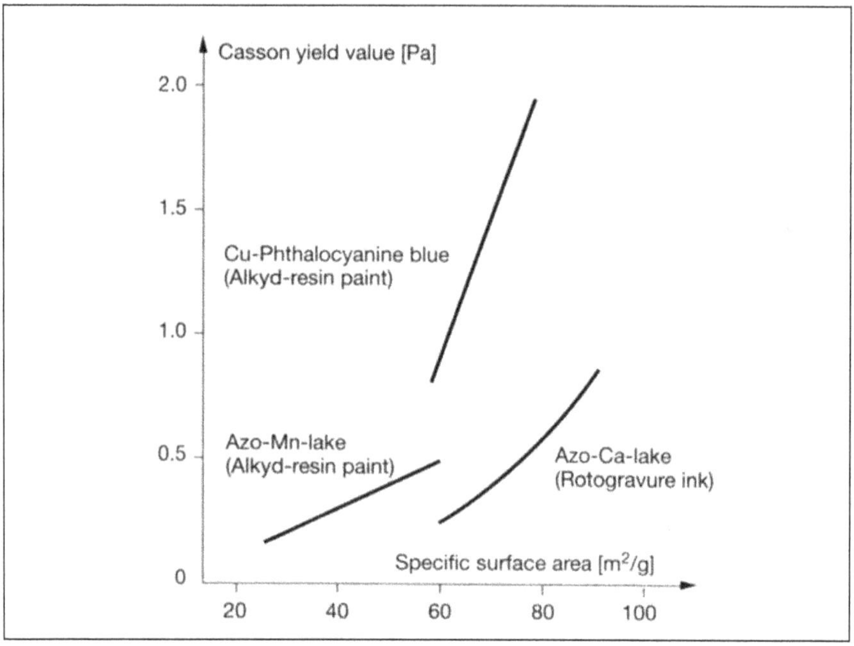

Figure 3.22: Examples for the dependence of yield value upon specific surface area
Source: According to K. Apel [121]

Surface character governs the interaction between pigment particles and the surrounding medium, and thereby the degree of adsorption on the pigment surface. A stronger interaction implies a lesser mobility of the particles and, consequently, an increase in viscosity.

With regard to particle shape, isometric pigment particles have a more favourable effect on the flow behaviour of pigment dispersions than anisometric ones. Thus, for example, under shear strain acicular pigments, like polymer chains, align in the direction of motion, thereby giving rise to pseudoplastic flow. Plate-like and acicular pigments with polar surfaces are more prone to build up thixotropic structures. This is attributed to the anisotropic charge distribution, which favours coalescence at corners and edges, i.e. the formation of a structure similar to a house of cards.

Specific surface area and volume concentration are decisive for the volume of adsorbed binder, which is responsible for the behaviour of the system under low shear stress. A series of tests carried out by *Apel* with analogous pigments, i.e. same surface character and same particle shape, showed that a nearly linear relationship exists between the yield value according to *Casson* and the specific surface area (Figure 3.22).

From the above, it can be inferred that the volume of the binder layer adsorbed and immobilised on the pigment surface is proportional to the available surface. Since this volume must be added to the volume of the dispersed phase, the relationship between specific surface area and flow properties of the dispersion becomes clear [12].

On studying the flow behaviour of pigmented linseed oil varnishes, *Honigmann et al.* determined that, whereas the final viscosity is dependent upon the viscosity of the varnish and the pigment concentration, the yield value – in addition to these two factors – is also influenced by the particle size and by the interaction between the pigment and varnish; i.e. the yield value is pigment-specific. This can be accounted for by the fact that the interaction forces between pigment and binder, as well as the mutual interaction forces between the pigment particles, are practically ineffective in the presence of high shear forces and thus have no effect on the final viscosity [48].

3.5.2 Printing inks

With the exception of screen printing, the thickness of print films is usually of the order of a few micrometres and thus the concentration of the colouring component in a printing ink must be much higher than in paints, enamels or plastics (see Table 3.7). For the same reason, the oversized particles in a printing ink should themselves not be larger than very few micrometres as otherwise poor gloss will result. Both factors considerably increase the influence of the pigment on the flow properties of a printing ink. The extent of a pigment's influence on the rheological behaviour of the medium is thus a determining factor for the application properties of the ink, and therefore a decisive criterion in the choice of pigments for printing ink media.

In view of such concentration ranges, it is understandable that extremely high rheological demands must be met in the case of printing inks. Under application conditions, solvent-containing inks, i.e. for rotogravure and flexography, with pigment concentrations at the lower limit of the ranges quoted in Table 3.7, can behave practically as Newtonian systems. Printing inks are, however, generally pseudoplastic, linear plastic, non-linear plastic or thixotropic, whereby these anomalies may also occur simultaneously. In many cases they are also accompanied by

Table 3.7: Usual pigment concentration ranges in printing inks for different printing processes (using almost exclusively finely-divided organic pigments)

Printing process	Amount of pigment [%]
Rotogravure	10 to 15
Flexography	10 to 20
Letterpress, offset printing	15 to 30
Lithography	20 to 35

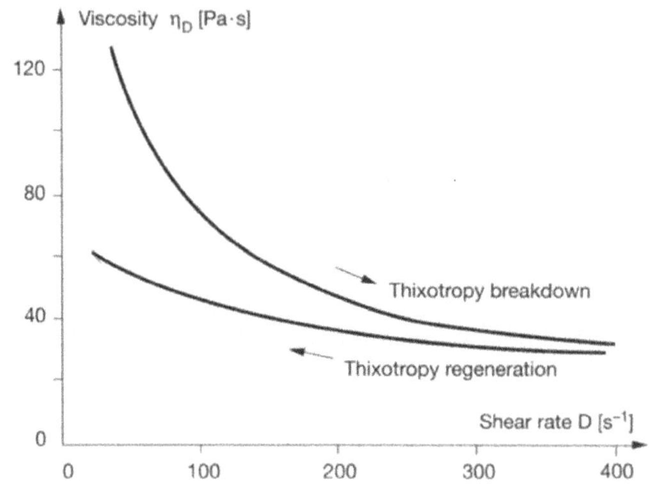

Figure 3.23: Viscosity curve of a quinacridone pigment (C.I. Pigment Violet 19) in an offset-printing ink before and after collapse of the gel structure

dilatancy and viscoelasticity. Figure 3.23 shows by way of an example the viscosity curve for a commercial offset printing varnish containing an organic pigment (viscosity of the pure varnish is approximately 8 Pa·s).

According to *Zettlemoyer* [63, 65], the application of printing inks may be divided into four phases as follows:

1. Fountain phase
2. Distribution on the cylinders
3. Transfer phase to printing roll or plate, and printing
4. Penetration into the printed paper.

The flow behaviour at low shearing forces is important for phases 1 and 4, whereas the rheological properties at high shear stress are decisive for phases 2 and 3. It thus follows that a high yield value will impair transfer of the ink from the fountain onto the cylinders via the fountain roller and its penetration into the printed paper. In rotogravure, the high yield value will also prove a disadvantage during the printing phase as, in this case, the ink will be unwilling to leave the recesses etched in the printing cylinder. On the other hand, when the final viscosity is high, this will affect transfer and distribution on the rolls as well as the printing phase itself.

Technical characteristics for printing

Most of the pigmented printing inks are sufficiently described by the rheograms, already discussed in Section 5.6 in connection with the plotting of flow anomalies, as well as by the rheological magnitudes **yield value** and **final viscosity**. For characterisation of the application properties, it is also possible to resort to two additional magnitudes proposed by *Zettlemoyer*, namely **shortness** and **tack** [65]. Shortness is the ratio of the yield value to the final viscosity of the ink. Tack is the reciprocal value of the shortness. Because accurate and reproducible measurements are difficult and troublesome to obtain in practice, the printing ink industry uses specially-designed measuring devices in order to assess shortness and tack. These instruments, termed **tackmeters**, simulate the application conditions occurring during the printing process. As a rule, the values thus determined are good leads for the practice. They are, however, only valid for the system examined and cannot be applied to other pigments or binders.

The shortness should be neither too high nor too low. If it is too high, corresponding to a very high yield value or a very low final viscosity, the ink will not follow the duct roll properly and its cohesion will not be sufficient for a uniform effect on the form rolls. Low shortness corresponds to a high tack. This concept is a measure for the ink's resistance to rapid split. Too great a tack of the ink gives rise to poor splitting of the ink film and tearing of the paper. Table 3.8 lists yield values for several printing inks which were produced with the offset varnish mentioned above and various organic pigments. The pigment concentrations quoted gave, at a shear rate of 400 s^{-1} after thixotropy breakdown, a viscosity of about 30 Pa·s. The history of the inks and the measuring conditions were kept strictly constant.

According to investigations performed by *Pahlke*, both thixotropy and viscoelasticity are strongly dependent on the polar forces between pigment particles and varnish. The formation of a frame structure of particles, i.e. flocculates, in the dispersion is attributed to interparticle attractive forces, indicating incomplete wetting or poor enclosure of the aggregates. When polar forces between pigment and binder are stronger than the mutual attractive forces of the pigment particles, the formation of a pigment structure is not or to only a limited degree possible. Thus, strong polar interaction between varnish and pigment results in a slight

Table 3.8: Yield values for some organic pigments in a printing varnish. The measurements were performed with a particularly sensitive viscometer

Source: Organische Pigmente für die graphische Industrie, Hoechst AG, 1964

Pigment type	C.I. Denomination	Concentration [%]	Yield value [Pa]
Dioxazine violet	Pigment Violet 23	18	> 2200
Quinacridone	Pigment Violet 19	18	1920
Cu-phthalocyanine	Pigment Blue 15	22	1600
Hansa yellow	Pigment Yellow 1	29	820
BONS, Ca-lake	Pigment Red 57	15	0
Thioindigo	Pigment Red 88	31	1000

thixotropy and a lower yield value. As to viscoelasticity, strong polar forces between pigment and varnish give rise to longer relaxation times. Viscoelastic properties make themselves particularly felt in ink distribution, transfer and splitting processes [64, 66].

3.5.3 Paints and enamels

As a rule, paints and enamels are plastic or pseudoplastic, anomalies which are often accompanied by thixotropy. Under extremely high shearing conditions, however, their flow corresponds rather to that of Newtonian fluids. As with shear rate, their viscosity also shows a wide range, lying between approximately 10^0 and 10^4 Pa·s. The rheogram in Figure 3.24 illustrates the relationship between viscosity and the shear rates occurring during storage and application of these systems.

Process-technical assignment

In the paper by *Pierce* [62] mentioned above, a survey is given of the processes involved in the manufacture and application of paints and enamels. The corresponding ranges of shear are stated and a discussion of important rheological aspects is included.

For all processes taking place in the low shear stress region, e.g. sagging, levelling, settling, the rheological behaviour is more conveniently characterised by the yield value. It is understandable that a high yield value is unfavourable for sag-

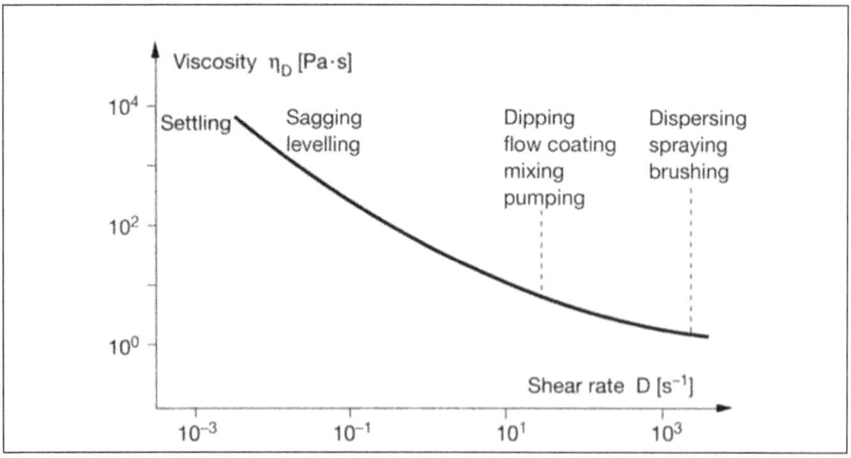

Figure 3.24: Viscosity of paints in dependence upon the shear rates occurring during their manufacture, storage and application

Source: NL Rheologie Handbuch, Kronos Titan GmbH, Leverkusen, 1980 ging and levelling. With regard to settling, a high yield value is, on the contrary, most desirable as it counteracts particle sedimentation. A high yield value also avoids sagging of applied paint films to take place. For this reason, **rheological additives** are sometimes added to paints and enamels during their manufacture. These **anti-settling agents**, as they are also called, build up colloidal structures and, due to the resulting yield value, retard or even eliminate settling. Such additives are, as a rule, **organic-modified layer silicates** or organic compounds, such as **synthetic waxes** and **castor oil derivatives**, whose particles build up gel-like structures after dispersion due to solvation and swelling [67].

The final viscosity is the more instructive parameter for those application methods using high shear forces, e.g. spraying and rolling. It should be low, because a final viscosity that is too high is definitely detrimental to this kind of process.

Influence of specific surface area

The relationship between specific surface area and flow behaviour is usefully employed to manufacture organic pigments for the paint industry with high hiding power that can be considered as possible substitutes for lead-containing inorganic pigments, i.e. lead chromates and lead molybdates. By appropriately controlling production conditions and finish processes, the particle size can be optimised on the one hand with a view to scattering (see Section 2.2) and, on the other hand, the growth of particles in the fine region of the particle size spectrum can be inhibited. The resulting organic pigments have high hiding power and low specific surface area, and their improved flow properties allow higher concentrations to be used in the paint system without impairing gloss [68].

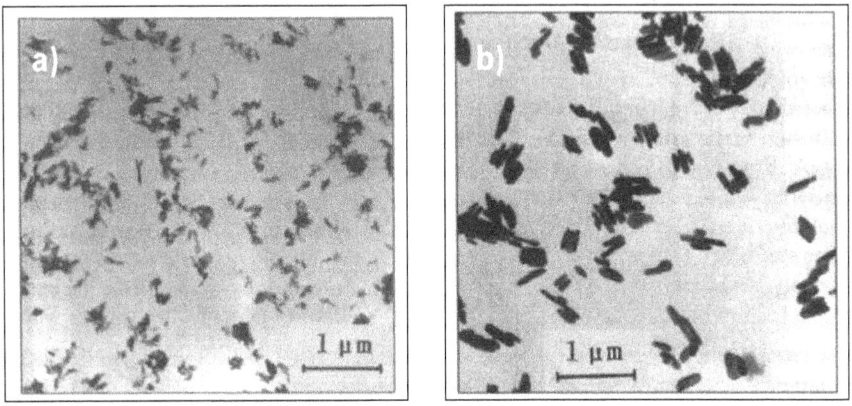

Figure 3.25: Electron photomicrograph of a disazo yellow pigment (C.I. Pigment Yellow 83, N° 21 108), a) standard grade b) grade with optimised hiding power

Source: Works micrograph Hoechst AG

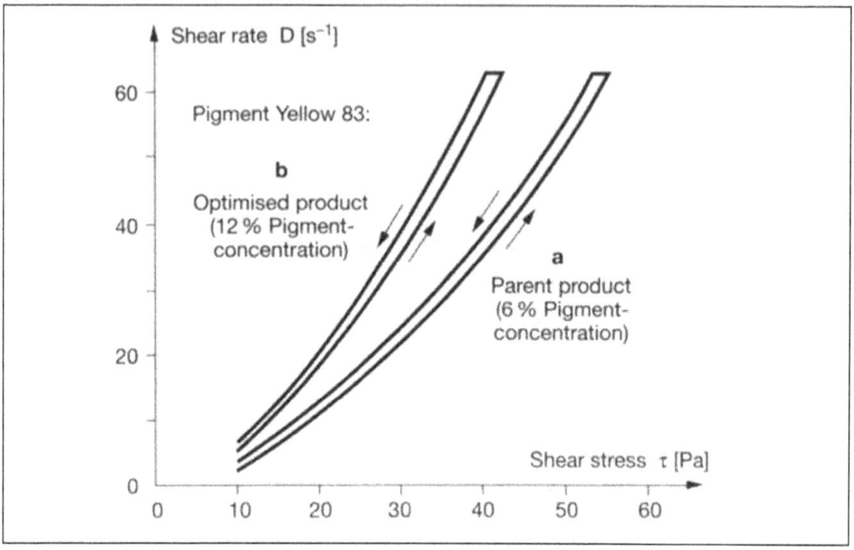

Figure 3.26: Flow curves of Pigment Yellow 83, standard and optimised grades, in a paint medium. The system with 12 % optimised grade shows a better flow behaviour than the system with 6 % standard grade)

For example, C.I. Pigment Yellow 83, N° 21108, a disazo pigment without lake-forming groups, may be quoted. The electron micrographs in Figure 3.25a and b show the distinct difference in particle size between the parent pigment and the grade with optimised hiding power. The corresponding flow curves are illustrated in Figure 3.26. Despite its higher concentration (12 % instead of 6 % in the case of the parent pigment), the optimised grade shows a much better flow behaviour.

The viscoelastic properties of paint layers or enamel films are assumed to be mainly determined by the structure of the high polymer, i.e. by the ratio of its crystalline to amorphous zones, and by the degree of cross-linking. There are, however, some controversies in the experts' opinion about the true facts. Since a close relationship exists between viscoelastic and mechanical features, the visco-elastic characterisation of a coat of paint or enamel allows conclusions to be drawn regarding its mechanical properties. A paper by *Molenaar* and *Hoeflack* contains a compilation of suitable methods for this purpose [69].

In paints, the increase in viscosity known as **thickening** can also be caused by the pigments. Apart from the thickening resulting from the premature oxidation of air-drying paint systems, basic pigments, e.g. zinc oxide, white or red lead, may give rise to thickening by reacting with acid (e.g. -COOH) groups in the binder. These two kinds of thickening are named **gelling** or **gelation**, and render the paint unusable. The phenomenon occurring in paint systems known as **floccula-**

tion also gives rise to thickening. It results in clusters of insufficiently stabilised pigment particles, thereby increasing yield value and thixotropy. In contrast to the thickening caused by oxidation or saponification, this fault is reversible (more information on flocculation is given in Section 4.3.1).

A study by *Zorll* discusses the flow properties of paint systems, their causes and methods for their characterisation [70]. A paper written by *Patton* deals with the physical fundamentals of the rheology of coating materials [71].

3.5.4 Plastics

Mass-coloration of plastics and rubber is generally carried out with far lower pigment concentrations than is the case for paints and printing inks, and the pigments' influence on the flow properties of plastic melts and rubber is in most cases negligible. However, carbon blacks and some organic pigments with extremely high particle fineness are exceptions, as in their case, due to their high specific surface area, even small amounts are sufficient to cause a yield value in the polymer matrix [72].

Master-batches and pigment preparations for plastics also constitute a particular case. Depending on pigment and medium, their pigment contents vary between 20 and 80 %, and thus carrier materials with low viscosity are generally employed for this purpose. Pigment preparations for the production of **thin foils** by extrusion blow moulding or for the manufacture of **staple fibre** are particularly critical, because an extremely fine dispersion state is demanded in these cases. Any agglomerates still present should not exceed a size of two to three µm, otherwise breakage of the filament can occur. On the other hand, the very fine dispersion state involves a high specific surface area and therefore a higher interaction between pigment particles and carrier material, and this may give rise to more pronounced flow anomalies.

Difficulties may also arise if the differences between the viscosities of the material to be coloured and the pigment preparation are too large. **Plastisols**, for example, have a relatively low viscosity as this is important for trouble-free application. As homogeneous coloration with powder pigments is not possible in their case, **pigmented platisciser pastes** are preferably used for this purpose. These pastes contain between 15 and 75 % of pigment, and are incorporated into the plastisol by stirring. If the aim is a fairly homogeneous final product, the viscosity difference may demand long stirring times. To avoid this problem, the amount of pigment/plasticiser paste needed for one production batch should first be mixed with an equal amount of the plastisol and only then, after due homogenisation, be added to the plastisol batch. Due to the lower pigment content, the flow properties of this diluted pigment/plasticiser paste are better adjusted to those of the plastisol to be coloured and shorter stirring times are the result.

3.6 Closing remarks

The reader is referred to the pertinent text books for more information on viscometrical methods. An excellent description of the viscometer types in use in the pigment processing industry is given by *Patton* [73]. For study of the flow anomalies often displayed by pigmented systems, viscometers allowing measurements to be made under varying shear conditions are suitable. Examples are the falling rod viscometer of the *Laray* type (DIN 53222, ISO 12 644) for heavy-bodied pastes (e.g. varnish printing inks), or the rotational viscometer with calibrated gap clearance. The latter is suitable for examining fluids as well as pastes. Nowadays an ample selection of commercial rotational viscometers is available. Figure 3.27 shows, for example, the "Universal Dynamic Spectrometer MCR 302" viscometer from Anton Paar Germany GmbH.

Figure 3.27: Rotating viscometer MCR 302 with computer for programmed measurement and evaluation

Source: Anton Paar, Germany GmbH

A comprehensive approach to the principles of viscosity measurement with rotational viscometers, including physical fundamentals and the pertinent mathematical derivations, can be found in booklets from *Anton Paar GmbH* [74] and *Thermo Fischer Scientific* (form. *Gebr. Haake GmbH)* [75]. Measurement procedures with these viscometer types are described in the DIN 53019-1 and DIN EN ISO 3219 standards.

4 Dispersion process: physico-chemical fundamentals

4.1 Introduction

The pigment dispersion process is of particular importance in the manufacture of materials coloured with pigments, e.g. paints, printing inks and plastics, as important properties of the final product, and not only those involved with colour, depend upon the particle size. In addition, the economic aspects of the dispersion process have a decisive influence on manufacturing costs and, thus, on the price of the finished material. Pigment producers, pigment processors and competent institutes have taken these facts into account over the last decades, and the results of their research have given rise to a scientific approach to pigment dispersion which is well-founded upon physico-chemical principles. This approach has superseded the former, generally empirical methods. Due to the rapidly changing processes, machinery and raw materials in this sector, this development will continue.

Pigments supplied in the form of powders consist of larger and smaller agglomerates. In order to thoroughly develop important application characteristics, such as tinting strength, hue, semi-transparency, gloss, hiding power, light and weather fastness, the pigment processor must reduce these agglomerates to their structural elements, i.e. aggregates and primary particles. The ideal case would be a complete reduction of the agglomerates to the original crystals, but this, if at all possible, would be far too expensive to be worthwhile.

The dispersion process is, as a general rule, not a true comminution process, i.e. breakage of the crystals does not occur. Crystal fracture can occasionally occur when pigments sensitive to grinding, such as plate-like or acicular pigments, are submitted to extremely high impact or shear forces. This phenomenon, known as **overgrinding** due to its detrimental effects on fastness and colour properties, is not only unwelcome, but should be avoided at all costs. On the other hand, incorrect dispersion may damage the surface of coated, i.e. treated, pigments. Research by *Schäfer*, for example, showed that the light fastness and stability to SO_2 of stabilised molybdate-red pigments may be impaired if the oxidic, stabilising surface coverage is damaged by extreme mechanical stress in the dispersing machine [76].

Juan M. Oyarzún: Pigment Processing

ISBN 978-3-86630-665-3

In the case of these pigments, extreme comminution also brings about a partial transmutation in the crystal modification. A decrease in hiding power and a shade change to a more yellowish red are the consequences.

4.2 Stages of the dispersion process

Pigment dispersion refers to a stepwise process whose objective is to produce a stable and uniform dispersion of finely-divided pigment particles, i.e. aggregates and primary particles, in an application medium, e.g. litho varnish, molten plastic or a synthetic resin solution. The sub-process termed **mechanical breakdown** comprises the disruption of the more or less weakly bound agglomerates into their constituent elements, i.e. aggregates and primary particles, by the action of mechanical forces. A further stage in the dispersion process is the **wetting** of the separated particles, whereby components of the medium spread over the particle surface to replace the initial pigment/air or pigment/moisture interface by a new pigment/medium interface.

Both sub-processes, mechanical breakdown and wetting, are followed by a third stage: stabilisation. This comprises uniform distribution of the separated and wetted particles and maintenance of the homogeneous, finely-dispersed state already achieved. In other words, a coalescence or re-agglomeration of the particles is to be hindered. This is attained by either the presence of an **adsorbed layer** of *vehicle molecules* on the pigment surface, or by the formation of an *electric double layer* around the particles. In this context, **re-agglomeration** refers to the renewed clumping together of non-stabilised aggregates and primary particles during dispersion.

In the course of the grinding operation, these three sub-processes do not necessarily take place in chronological order, but rather occur partly successively and partly simultaneously. It is conceivable that the agglomerates are first wetted and then, due to weakening of the attractive forces, disrupted. Other particles, however, if the shear and impact forces are sufficiently high, may first be moistened, i.e. incompletely wetted (initial wetting), thereafter disrupted and only then completely wetted (intimate wetting). Whichever is the case, uniform distribution and stabilisation are only possible after disruption and wetting. These stages will be discussed separately in the following sections.

4.2.1 Mechanical breakdown

4.2.1.1 Forces of interaction between the pigment particles

The atoms, ions or molecules of solids are held together by the action of their intermolecular forces of attraction, the **cohesion forces**. We know from experience that the cohesion forces must be overcome to smash or pulverise a solid body, and that this generally requires a considerable expenditure of energy. In this context, the following quantity relationships should be kept in mind: the mass of particles decrease with the third power of their diameter, whereas the attractive forces between them decrease approximately with the second power of their diameters. Thus, the attractive forces between particles with microscopic dimensions, as they occur in colloidal disperse systems, are considerably higher than in case of coarse-grained substances, e.g. sand. In the range of 0.01 to 1 mm, the weight and attractive forces of a particle may be of the same magnitude. In particles with a diameter of 1 μm, however, the force of attraction may amount to a million times its weight. This explains the increasing tendency to re-agglomeration with decrease in particle size.

The cohesion forces acting between the structure units of pigment agglomerates are essentially physical in nature. They differ from the forces of chemical bonding in that a transfer of electrons between the atoms does not take place. They are composed of:

- **van der Waals attractive forces**
These are electrostatic in nature and based on the interaction between permanent dipoles as well as on attractive forces arising from dipole/dipole orientation or induction effects. Orientation effects are termed **Keesom effects** and induction **Debye effects** (after their discoverers). Due to the fact that their interaction energy is proportional to the inverse sixth-power of the distance, the range of action of the van der Waals forces diminishes rapidly with increasing separation. Molecular crystals are held together by van der Waals forces, and thus they occur in organic pigments.

- **London forces**
London forces, often also termed **dispersion forces**[1], originate in particular in dipole-free molecules and atoms and are attributed to rapidly varying distributions of charge. These arise due to fluctuations in the electron density and give rise to the formation of transient dipoles. These transient dipoles, in their turn, induce instantaneous dipoles in the neighbouring molecules, whereby instantaneous interaction with the original dipoles is brought into effect. Hence, dispersion forces are, like the van der Waals forces, electrostatic in nature and their range of action is extremely short. They are usually considered as a special case of the van der Waals forces. In non-polar molecules, they are the sole cause of van der Waals attraction. In molecules with a dipole moment of up to 1 Debye, the dispersion effect also outweighs the orientation effects. The

[1] These forces, although discussed here in context with pigment dispersion, have nothing to do with that forces. The origin of the term can be found in the field of optics, where also "dispersion" is of importance; compare the remarks Section 2.3.1 footnote on page 71.

London forces increase quickly with growing molecular volume, on account of the higher polarisability. For this reason, the electrical polarisability of a molecule is a measure for its ability to exert dispersion forces. They occur primarily in non-polar organic pigments.

- **Coulombic forces**
  Electrostatic attraction in its proper meaning is based on permanent charges in the molecule. It results, for example, in the ionic crystals of salts and oxides, and therefore mainly in inorganic pigments like titanium dioxide, mixed-phase pigments, iron oxides and substances of similar structure. Coulombic forces are, as a rule, weaker than van der Waals or London forces, but they even attain the magnitude of covalent bonds when acting between ions. Moreover, Coulombic forces are, because of their inverse 2nd power dependence on the distance, considered as long-range forces.

- **Hydrogen bonding**
  This type of interaction occurs in groups in which hydrogen is linked to highly electronegative atoms, e.g. O, N, F. The hydrogen atom is thus positively polarised and has a strongly attractive influence on neighbouring negative ions or negatively polarised atoms. Hydrogen bonding occurs in pigments containing hydrophilic groups such as –OH, $-NH_2$ and -COOH. The resulting attractive forces are appreciably higher than van der Waals or London forces.

From the above list it can be concluded that, except for the London forces, all other interactions are substance-specific. The London forces presuppose only the presence of electrons and are thus effective in all types of substances. Their strength is governed solely, as already mentioned, by the polarisability of the individual molecule. A characteristic of van der Waals and London forces is that they are only effective over very short distances, and roughness on the crystal surface reduces their effects. This fact is sometimes used in practice to lessen the tendency to agglomeration. Attraction forces between dipoles are proportional to r^6, where 'r' is the distance between two particles. Electrostatic forces are less dependent on surface roughness [77].

Although chemical constitution is responsible for the **strength** of the cohesion forces; their **effectiveness** is governed by several additional factors like particle shape, average particle size and particle size distribution. The specific surface area depends on the average particle size. High specific surface area, occurring in finely-divided organic pigments, implies a higher extent of effective surface energy. The number of contact sites is determined by particle shape and particle size distribution.

The models shown in Figure 4.1 demonstrate the influence of particle shape on the number and extent of contact sites. Particles that are approximately spherical (i.e.

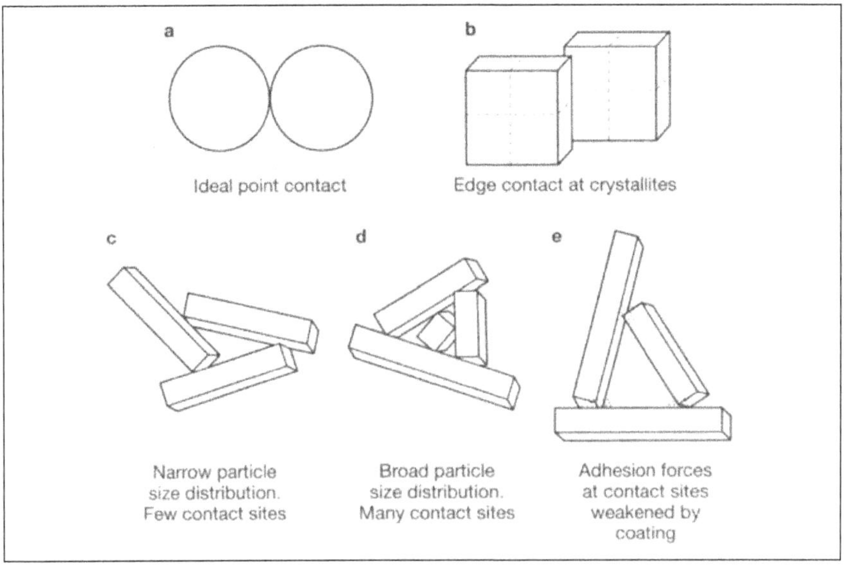

Figure 4.1: Influence of shape and distribution width of pigment particles on number and extent of contact sites (see text for further explanations)

nodular particles) can only coalesce in practice via point-like contact sites (Figure 4.1a). Conversely, in the case of agglomerates formed by plate-like or cube-shaped aggregates, contact along the edges is possible (Figure 4.1b). Moreover, in anisometric particles the different chemical constitution of the crystal surfaces may give rise to an additional polarity. Figures 4.1c and 4.1d show the effect of particle size distribution. A wide distribution range brings about an increase in contact sites. At the same time, small particles can block the gaps and cavities between the larger particles, and inhibit penetration by the medium and, thus, the wetting. The special case of a surface-treated pigment is shown in Figure 4.1e. As the coating agent shields the cohesion forces, the effect of the contact site is weakened. Such pigments can be easily dispersed.

4.2.1.2 Types of mechanical breakdown

The resistance of agglomerates against reduction to their original structural units was expressed formerly by the concept of grain hardness. This expression was not a particularly good choice, because it could easily be confused with the physical hardness of crystals, i.e. the resistance of crystals of a specific solid to scratching by another solid. The hardness of a crystal is a material property, which is generally measured by means of the empirical hardness scale according to Mohs. The term grain hardness was later replaced by the more appropriate expression dispersion hardness, introduced with DIN EN ISO 8781-1 in order to specify a num-

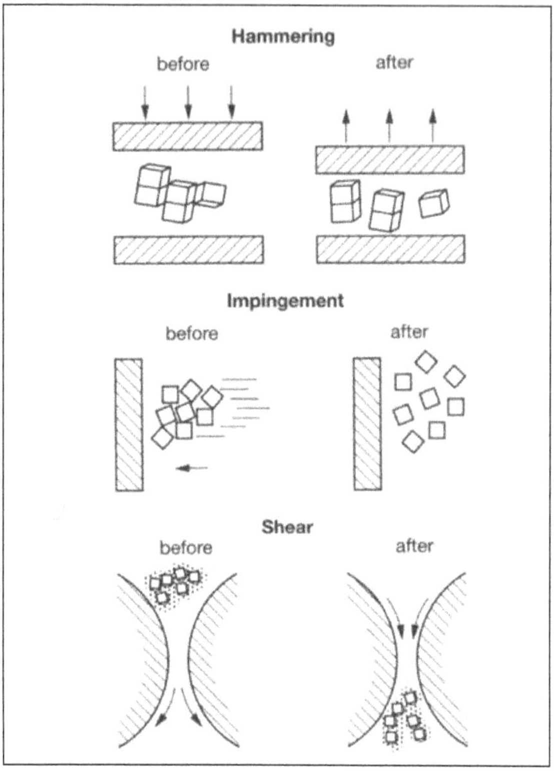

Figure 4.2: Types of mechanical breakdown of pigment particles

ber characterising the amount of work that is required in a dispersion to develop the tinting strength of a coloured pigment.

In this context, the discussion of hardness becomes questionable. For binder systems the concept of dispersion hardness has nowadays been substituted by the more adequate expression **ease of dispersion,** as this is the final point of the matter. According to DIN EN ISO 8780-1, the ease of dispersion is a measure of the rate at which a pigment achieves a given level of dispersion during its milling in a binder system. The term **dispersion hardness** is still valid for testing the dispersibility of pigments in a plasticized PVC on a three-roll-mill (DIN EN ISO 13900-2).

Unlike crystal hardness, the ease of dispersion is not a specific property of a pigment, as its density or refractive index, but is also governed by the conditions of dispersion, i.e. medium, grinding machine, temperature etc.

The break-up of agglomerates into their structural units is accomplished, acco-ording to *Patton* [78], by smashing, impingement or shear (schematically illustrated in Figure 4.2).

Smashing takes place when pigment particles are caught between two solid surfaces, e.g. colliding balls in a ball mill or pestle and wall in an edge-runner in the course of dry-milling. Break-up by **impingement** is carried out when particles are flung against each other, against grinding bodies or the wall of the dispersing machine. A good example of this is the jet-mill, where the particles, by means of a stream of compressed air or superheated steam under high pressure, disintegrate by impinging against one another. Division by **shear** is accomplished by shear forces, resulting, for example, in a printing varnish between the rolls of a triple-roll mill.

It is apparent that these types of breakdown demand different viscosity conditions. Whereas smashing and impingement are only possible in systems of low viscosity, the transfer of high shearing forces presupposes a high viscosity of the mill base. Doubtless in some dispersing machines, such as ball and bead mills, all three types may occur. This is not the case for heavy-duty mixers, triple-roll mills and high-speed disc dispersers, where dispersion is carried out solely by the action of shear forces.

Patton's conceptions about the types of breakdown resulting in grinding processes are, however, not completely shared by other experts. Thus, *Rumpf*[79] considers four different types:

1. Strain between two solid surfaces, e.g. smashing as in the course of dry-milling in an edge-runner.
2. Strain at one solid surface, e.g. strain by impingement, as occurring in jet-mills.
3. Strain by shear in the surrounding medium (wet-milling in ball, bead and triple-roll mills).
4. Cracking by thermal strain.

In his opinion, the break-down of agglomerates in ball and bead mills may in the main also be attributed to shear forces, and not to contact with grinding bodies or other particles. The underlying assumption here is that true smashing and impingement in wet milling are not possible on account of the medium present, and any such effects would also make themselves felt in the form of shearing forces.

4.2.2 Wetting

4.2.2.1 Energetic approach

Wetting is understood to be the formation of an intimate contact surface between pigment and medium. In order to obtain this intimate contact, the extraneous molecules attached to the pigment surface (air, moisture or other contaminants) must be replaced by molecules of the continuous phase, i.e. the **pigment/air** or **pigment/moisture** interface must be replaced by the **pigment/medium** interface. The adsorption energy evolved by this process, the **wetting energy**, diminishes by the amount of energy required to displace the extraneous molecules.

The wetting energy is a measure for the interaction between pigment and medium. The higher the wetting energy, the better the wetting. The highest wetting energy results in the case when the bonding forces attaching the medium molecules to the pigment surface are chemisorptive in nature. The cohesive forces in the medium and the adhesion forces between medium and pigment surface are decisive for the interaction between pigment and medium. When the adhesive forces are higher than the cohesion forces, a strong interaction takes place and, consequently, the pigment can readily be wetted. In this case, the pigment surface is characterised by the term **lyophilic** (solvent-attracting). In the opposite case, i.e. if the forces of cohesion of the medium molecules are higher than the adhesive forces, wetting does not occur spontaneously and a higher expenditure of dispersion energy is required to bring it about; the pigment exhibits a **lyophobic** (solvent-repelling) behaviour. These facts may well be expressed by saying: *"In the first case, the pigment is at ease with the medium and willing to accept relations with it. In the second case, the pigment feels ill at ease in its surroundings and tries to keep to itself. A relationship between both is only possible under compulsion."*

This pigment behaviour, differing according to the medium in which the pigment is dispersed, can be described from an energetic point of view. The concept of **spreading**, i.e. the spontaneous flowing of a liquid on a solid surface was introduced in Section 1.4.2. Since the wetting process arises from competitive action between adhesive and cohesive forces, the rate of wetting may be characterised by the balance of the work of adhesion and cohesion after equilibrium is established. For this reason, the name **spreading coefficient** K_{spr} is given to the difference between the work of adhesion and cohesion.

Equation 4.36:

$$K_{spr} = A_{ad} - A_{co}$$

By definition, the work of adhesion is the work necessary to separate 1 cm^2 of the interface and to re-form the two original phase surfaces, in our case solid and liquid. Thus, on the one hand, the expenditure of work is equal to the interface energy γ_{sl}, and, on the other hand, the surface energies σ_s (solid) and σ_l (liquid) are recovered. Therefore, the work of adhesion is given by the following equation obtained by *Dupré*:

Equation 4.37:

$$A_{ad} = \sigma_s + \sigma_l - \gamma_{sl}$$

The work of cohesion required to separate 1 cm^2 of the liquid surface must obviously be equal to $2\sigma_l$, because the separation gives rise to two 1 cm^2 large liquid surfaces. By introducing these expressions for the work of adhesion and cohesion in Equation 4.36, the resulting energy balance is:

$$K_{spr} = \sigma_s + \sigma_l - \gamma_{sl} - 2\sigma_l$$

$$K_{spr} = \sigma_s - \sigma_l - \gamma_{sl}$$

According to Young's equation (Equation 1.7), the difference $\sigma_s - \gamma_{sl}$, i.e. the wetting tension, is equal to $\sigma_l \cdot \cos\omega$, and thus the spreading coefficient can also be written:

Equation 4.38:

$$K_{spr} = \sigma_l (\cos\omega - 1)$$

Spreading occurs if, as discussed above, the work of adhesion is higher than the work of cohesion, i.e. if $K_{spr} > 0$, and this is true only when $\omega = 0°$, as for $\omega > 0°$ the quantity K_{spr} must be negative. Incidentally, Young's equation is not valid for $\omega = 0°$. Here, $\cos\omega = 1$ and the wetting process takes place spontaneously. Equation 4.38 is, therefore, valid only for cases where spreading does not occur.

Both quantities, the surface tension of the liquid σ_l and the contact angle ω resulting from partial wetting, can be measured. Hence it follows that Equation 4.38 allows a quantitative assessment of the spreading coefficient and, consequently, a numerical appraisal of the influence of temperature or additional solvents or surfactants on the spreading behaviour.

A further method to ascertain if a pigment is properly wetted or not by a given medium consists in assessment of the critical surface tension, as explained in Section 1.4.2.

However, a limitation must be imposed upon the energetic relationships just discussed. Strictly speaking, they apply only to absolutely smooth solid surfaces and are therefore not applicable to our case, namely pigment powders. For this reason, accurate calculations demand the introduction of a correction factor F_{Ru} (Ru for **rugosity**) for $\cos\omega$, which takes into account the solid surface enlarged by roughness. This correction factor is, as a rule, determined empirically and is always higher than 1. Consequently, Young's equation for rough solid surfaces becomes:

$$\sigma_s = \gamma_{sl} + F_{Ru} \times \sigma_l \cos\omega$$

where $F_{Ru} > 1$. The spreading coefficient thus becomes:

$$K_{spr} = \sigma_l (F_{Ru} \times \cos\omega - 1)$$

Summing up, it may be stated that for adequate wetting, the surface tension of the medium should be the lowest possible, as this gives rise to a smaller contact angle. Thus, low polymer concentration, and therefore low viscosity of the medium, have a favourable influence on the wetting process. A further means to decrease the surface tension of the medium is the addition of wetting agents [80, 81].

4.2.2.2 Wetting kinetics

Apart from the thermodynamic conditions that the achievement of wetting in a pigment/medium system presupposes, the **rate** at which wetting takes place in practice is also of interest. Pigment powders are usually added to the litho-varnish, binder solution or plastic dispersion under stirring, so that wetting starts before the proper grinding process is initiated. During premix of the mill base, i.e. the incorporation and uniform distribution of the pigment in the medium generally carried out in a high-speed disc disperser, the major part of the entrapped air is displaced and the air on the pigment surface replaced by the medium. Some air residues, however, may remain behind in the interstices of the agglomerates. During the subsequent penetration of the medium into the pores and capillaries, the agglomerates are cracked open by the increased pressure and the remaining air can escape.

This oversimplified, partial description of the very complex wetting process is justified as the rate of penetration of the medium into cracks, pores and capillaries of the agglomerates is likely to be decisive for the speed of the whole wetting phase.

The flow velocity of a fluid passing through the cross-section of a pipe is given by the *Hagen-Poiseuille's* law:

$$\frac{dV}{dt} = \frac{\pi R^4 \Delta p}{8\eta l}$$

This equation states that at a given pressure gradient, Δp, the differential volume dV of a liquid passing through a cylindrical pipe in a time differential dt is proportional to the fourth power of the inner radius, and inversely proportional to the fluid viscosity η and to the pipe length l. *Washburn* [82] replaced the pressure gradient Δp by the capillary pressure $2\sigma_l \cdot \cos \omega / R$ (Equation 1.10 taking the correction factor $\cos \omega$ into account), i.e. the driving force of the wetting process, and derived the following expression, named after him, for the velocity at the beginning of the wetting process for horizontal, cylindrical capillaries and incomplete wetting:

Equation 4.39:

$$\frac{dV}{dt} = \frac{\pi R^3 \sigma_l \cos \omega}{4 \eta l}$$

where

V = volume of the penetrating medium,
t = the time,
R = statistical mean of the capillary radii,
σ_l = surface tension of the medium,
ω = the contact angle,
η = the dynamic viscosity of the medium,
l = the length of the capillaries.

Thus, the rate of wetting during the initial stage is directly proportional to the surface tension and inversely proportional to the viscosity of the medium. Furthermore, the rate of wetting is strongly dependent upon the capillary radius. Naturally, substantially more complicated relationships are required for pigment powders but, nevertheless, the Washburn equation gives a reference point for the magnitude of the rate of wetting.

The pigment processor is now confronted with the problem that low surface tension of the medium is favourable for wetting, but unfavourable for the rate of wetting. Since good wetting is essential for a stable dispersion and, on the other hand, the rate of wetting can be regulated by varying viscosity and temperature, preference should be given to stabilisation, and therefore the surface tension of the medium ought to be adjusted as low as possible. The dependence of the rate of wetting upon the third power of the capillary radius indicates that pigments consisting of loosely formed agglomerates are more quickly wetted than pigment agglomerates with a high packing density. However, this is not generally of great interest to the pigment processor, as other pigment properties are of greater importance to him when making his selection,

Apart from the addition of wetting agents, the most important factor influencing the speed of moistening or wetting is the viscosity of the medium. In the case of paints and solvent-based printing inks, a low viscosity of the binder solution may be attained by reducing the binder concentration. However, this method has its limitations, because insufficient binder content in the medium promotes wetting and coating of the aggregates with solvent, and this should be avoided. Further means are an increase in temperature, in the case of closed grinding equipment, and an adequate choice of solvent or solvent mixture.

4.2.3 Stabilisation

4.2.3.1 Fundamental aspects

Colloidal dispersions, which also include pigment dispersions, are thermodynamically unstable. The energy input required for production results in surplus energy on the surface of the dispersed phase. Hence, the dispersed particles exhibit a tendency toward attainment of a state of lower energy, i.e. the more stable state of the compact phase. They are thus prone to clump together, releasing their surplus energy. At the beginning of this section (see Section 4.2), the clustering of particles known as **re-agglomeration** was mentioned. A further form is **flocculation**, which occurs when already dispersed pigment particles, i.e. chiefly aggregates and primary particles, congregate during a later stage of processing to larger, loose clusters, the **flocculates** or **flocs**. Flocculates are wetted by the medium and can be readily redispersed by low mechanical forces.

When, by virtue of Brownian motion, the separation between the finely dispersed particles becomes small enough to allow intermolecular attraction forces to come into effect, flocculation is programmed in advance. In practice, we nevertheless succeed in producing stable pigment dispersions, and it must be possible to create specific conditions within the pigmented system that obstruct or impede clustering of the particles. One such condition is, for example, a high viscosity as it counteracts particle migration. Thus, the high viscosity of molten plastics results in stabilisation problems in these systems being virtually unknown. On the other hand, in the case of low viscosity pigment dispersions like paints or solvent-based printing inks, stabilisation is attained by bringing about other conditions. These are the two mechanisms briefly mentioned at the beginning of this section:

1. Steric hindrance by adsorbed polymer chain
2. Electrostatic repulsion between particles of the same charge.

In order to obtain sufficient stabilising effect, the thickness of the adsorbed layer on the pigment surface must be at least 5 nm. Generally speaking, such layer thicknesses result solely from the adsorption of polymers, and thus low molecular substances are unable to build stabilising adsorption layers due to their small molecular size. As a rule of thumb, it should be noted that steric hindrance and electrostatic repulsion play the respective main roles in non-aqueous and waterborne systems.

4.2.3.2 Stabilisation by steric hindrance

The mechanism of stabilisation by polymer adsorption onto the particle surface is based on **osmotic, enthalpic** and **entropic** effects which give rise to a mutual repulsion of the adsorbed layers. It is also possible that the polymer shell may induce flocculation of the disperse system. This phenomenon, termed **bridging flocculation**, is particularly caused by high molecular weight polymers having a strong affinity for the pigment becoming adsorbed simultaneously on several particles. This happens especially when the high molecular weight polymer is present in a small concentration.

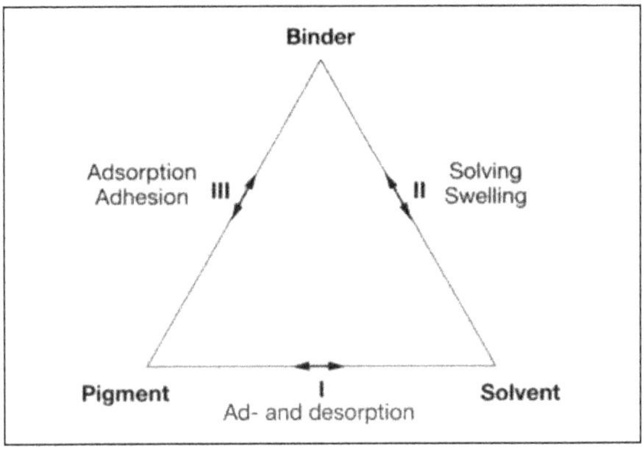

Figure 4.3: Possible interactions between the main components of a pigmented system (schematic)

The formation of adsorbed layers from polymer solutions on the surface of finely-dispersed pigment particles is a competitive process, as not only polymer chains, but also solvent molecules may be adsorbed. The triangular diagram in Figure 4.3 shows the possible interactions in a conventional, solvent-based printing ink or paint system. The triangle sides designated I, II and III represent the interactions between the main system components. These interactions are:

I The pigment/solvent interaction. It should be low, as otherwise solvent adsorption can take place

II The solvent/binder interaction. It should be middle-range. If it is too high, it favours the desorption of molecules already adsorbed. Conversely, polymer solutions containing 'bad' solvents facilitate polymer adsorption on the pigment surface.

III The pigment/binder interaction, which naturally should be as high as possible.

Pigment/binder interactions

Anchor groups and adsorption mechanisms

The interaction between the pigment surface and the active segments present in the macromolecules of the high polymer, the **anchor groups**, is caused by the intermolecular forces discussed in Section 4.2.1.1. In organic pigments, which as a general rule consist of molecular crystals, these forces comprise the relatively weak van der Waals and London forces. In the case of inorganic pigments, usually salts, oxides and sulphides of transition elements, the lattice energy of their ionic crystals is composed of Coulombic, van der Waals and London forces. In both cases, that is in organic as well as in inorganic pigments, hydrogen bond forces

Table 4.9: Magnitudes of the binding energy in kJ/mol for the different intermolecular forces

Intermolecular force	Bond energy [kJ/Mol]
Van der Waals forces	
a) permanent dipoles b) induced dipoles	< 20 < 2
Dispersion forces	< 40
Hydrogen bond	20 to 45
For comparison:	
Covalent bond	60 to 700
Ionic bond	600 to 1000

may also contribute to the interaction if the molecular structure is suitable. Table 4.9 gives an impression of the magnitudes of the intermolecular forces.

The anchor groups of the polymer molecules should be firmly attached to the pigment surface. The polar surface of inorganic pigments facilitates the formation of strong adsorption sites for the binder. This is not so easy with organic pigments due to their low polarity and often hydrophobic character. Pigment producers strive to counteract these facts by surface treating critical pigments, and have attained good results using for example amines, **rosin** or **rosin derivatives** as coating agents (see Section 1.4.4).

According to the model concepts of *Napper*, stabilised pigment particles are enveloped in a cloud of mobile polymer chains attached to the particle surface via their anchor groups [83]. The various possibilities for attachment and shapes of the layers adsorbed on the surface of the dispersed phase is shown in Figure 4.4.

The brush-like layer (Figure 4.4a) arises when the adsorption involves a single-point attachment and the rest of the chain projects as a thread into the solution. When the molecular mass is low and several anchor groups are present, i.e. in the case of high adsorption energy, the polymer chains will lay almost flat on the particle surface (Figure 4.4c). In such cases the adsorption process can be well represented by the Langmuir adsorption isotherm. If the energy of adsorption is not very high, the polymer shell is in the shape of either a coil (Figure 4.4b) or a loop (Figure 4.4d), depending on whether the adsorption involves one or several points of attachment. These configurations can be particularly expected when both molecular mass and concentration of the polymer solution are high.

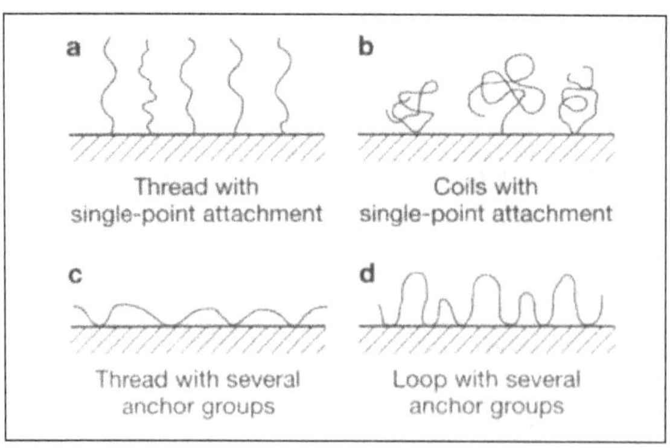

Figure 4.4: Models fir the possible configurations of adsorbed polymers layers on a pigment surface

As an example, the adsorbed layer of a linear polyester on a water-containing silicon dioxide is represented in Figure 4.5. In this case, the anchor groups are the carbonyl groups of the polyester chain which interact with the points of attachment on the SiO_2-surface, i.e. the hydroxyl radicals. Thus, adsorption takes place here by the formation of hydrogen bonding. The configuration corresponds to the scheme in Figure 4.4d and can be determined by IR-spectroscopic measurement [84].

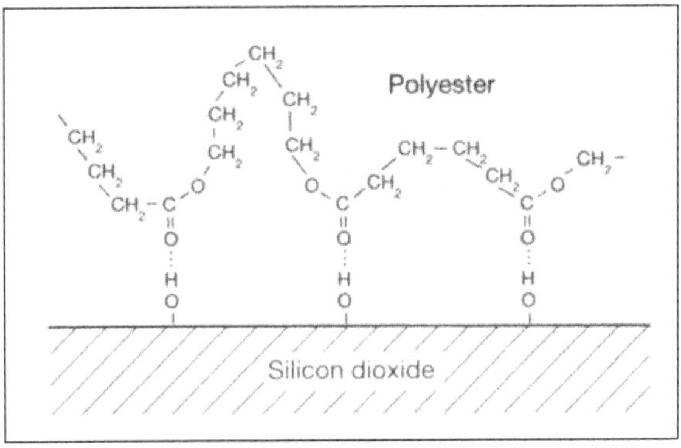

Figure 4.5: Adsorption of a linear polyester on a water-containing silicon dioxide. The anchor groups of the polyester form hydrogen bonds with the hydroxyl groups on the solid surface

Threads and coils with only a single-point attachment form thicker layers than threads and loops with several points of attachment. However, adsorption layers attached at several points are more stable, as they have less tendency to be desorbed due to their stronger attachment. Nevertheless, a higher number of points of attachment does not necessarily result in better stabilisation as the nature of the active groups and their distance apart on the pigment surface also have a decisive influence on bonding strength and the configuration of the polymer shell. *Napper* attributes the stabilising effect of the adsorbed layer to the freely moving segments of the macromolecules, which extend into the binder solution to form a kind of barrier around the pigment particle and thus prevent it from joining other particles (Figure 4.6). Block and graft polymers give rise to better barriers than homopolymers.

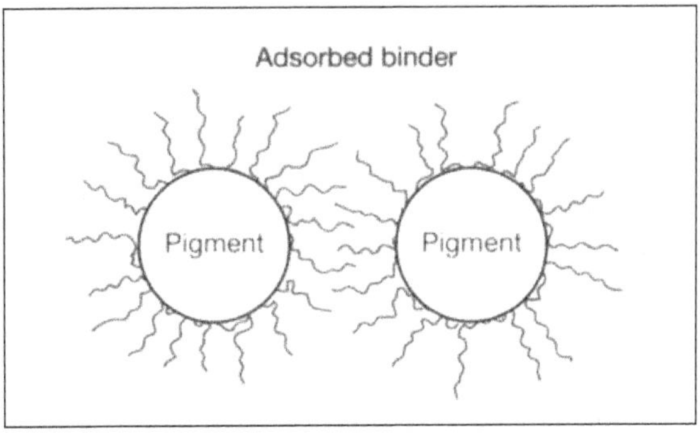

Figure 4.6: Stabilisation arising from the formation of a steric barrier by binder adsorption according to Napper

From these relationships, confirmed by the research of *Hamann* and *Joppien* [84, 85, 86], it can be deduced that several factors are decisive for the formation of a stabilising adsorbed polymer layer around the particles of the dispersed phase. These factors include, on the one hand, **chemical structure**, i.e. type and number of anchor groups in the macromolecule, mean molecular mass, molecular mass distribution and structure of the high polymer, and on the other hand **surface character of the pigment** and the **arrangement of the polymer chains** on the pigment surface. In addition, the adsorption process is also governed by temperature, type of solvent, additives and polymer concentration.

Thickness of the adsorbed layer

Several methods have been developed to determine the thickness of the adsorbed polymer shell. Starting from composites of adsorption isotherms, *Řeháček* found thicknesses of between 7 and 14 nm for polyester shells also containing solvent molecules on titanium dioxide and iron oxide pigments [87]. On the basis of viscosity measurements, *Rothstein* and later *Schaeffer* calculated the diameters of titanium dioxide and copper phthalocyanine blue particles dispersed in a polyester solution. For comparison purposes, the same pigments were also dispersed in the pure solvent. Layer thicknesses of the magnitude of 10 to 30 nm [88, 89] were determined. Also using a viscosimetric method, *Sato* found thicknesses of 2 to 20 nm for polyamide layers on carbon black, iron oxide and titanium dioxide [90].

Role of the acid-base properties

By characterising pigment and binder according to their acid-base properties, it is possible to draw conclusions as to their interaction. Depending on whether they behave as a proton donor, proton acceptor or otherwise, pigments as well as polymers can be classified in four groups: acidic, basic, amphiprotic and neutral. Pigments can best be dispersed in resins having the opposite acid-base character, i.e. basic pigments in acid resins and acid pigments in basic resins.

A thorough study on these relationships was carried out by Sørensen, who stated that PVC/PVA copolymers, polyacrylates and cellulose esters belong to the acid resins, whereas melamine, polyamide and alkyd resins belong to the basic ones. The acid-base character of pigments is strongly governed by their surface treatment. Thus titanium dioxide and carbon black, depending on the grade, may show acid or basic behaviour. Lead chromates and lead molybdates are generally amphiprotic. Arylamide yellow pigments are basic or neutral. Phthalocyanine blue grades stabilised against flocculation normally behave as proton donors, although the various types of this pigment may appear in all four groups. The above-mentioned study contains detailed information and tables [91]. The acid-base character can be determined by simple titration measurements.

Influence of the solvent

Competition between binder and solvent adsorption

The solvent governs the form and degree of solvation of the macromolecules in the solution, factors which in their turn have an influence on the configuration of the adsorbed layer. In addition, the solvent also has an effect on the rate of adsorption as, due to its smaller molecules, it can penetrate more rapidly into the pores and capillaries of the agglomerates and gain possession of the adsorption sites on the particle surface. In this manner, polymer adsorption is delayed. The macromolecules reach the adsorbent surface only later and must first displace the solvent molecules from the adsorption sites.

One refers in this context to **competitive adsorption**, and the affinity of the solvent for the pigment should therefore not be too high. On the other hand, the solvent must provide good solvation of the binder, i.e. solvent molecules ought to accumulate at the polymer chains, but again not be such a strong solvent that adsorbed macromolecules may be too easily desorbed from the pigment surface. *Dittrich*, *Oesterle* and *Hergesell* introduced the terms **effective wetting** and pseudo-wetting for binder and solvent adsorption respectively on the particle surface [92]. The pigment processor must therefore strive for effective wetting and avoid **pseudo-wetting**.

A study by *Carr* supplies us with a good example for the influence of the solvent or solvent blend on the stability of a pigment suspension. Dispersion tests carried out with a BONS-monoazo pigment (C.I. Pigment Red 57) in a gravure ink medium (an aromatic-soluble calcium-zinc resinate resin in mixtures of toluene and aliphatic petroleum distillate) showed a distinct improvement in the level of dispersion with decreasing toluene concentration in the aromatic/aliphatic solvent blend (Figure 4.7). Quite obviously, a high toluene concentration in the mixture impairs the polymer adsorption, thus reducing the stabi-

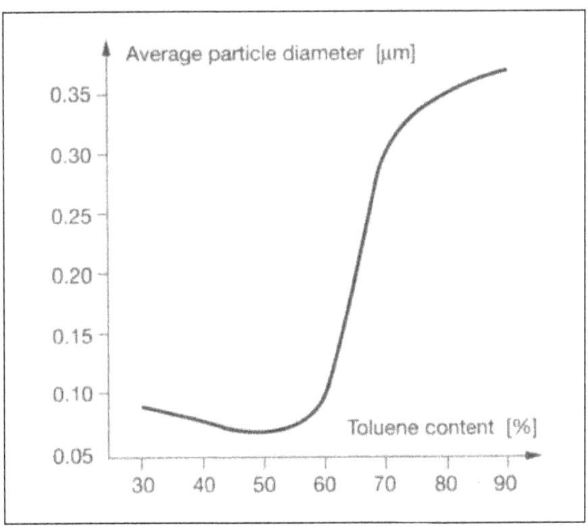

Figure 4.7: Particle diameter of a monoazo pigment as a function of the toluene content in the solvent mixture of a rotogravure printing ink

lising effect. The same series of experiments with a modified phenolic resin resulted in a consistently high level of dispersion, even for the test in pure toluene. It was thus concluded that in the case of high pigment affinity of the resin, the restraining effect of a strong solvent is negligible [93].

Solubility parameter

Relationships between resins and solvents can be characterised by means of the **solubility parameter**. The concept of this system, first introduced by *Hildebrand*, is derived as follows: The cohesive energy that holds together the molecules of a solvent or a binder consists of dispersion (*D*), polar (*P*), i.e. van der Waals forces, and hydrogen bonding forces (*H*). In order to overcome these forces, an amount of energy equal to the internal vaporisation energy E is required. Splitting this energy into its individual components, E_D, E_P and E_H, results in:

$$E = E_D + E_P + E_H$$

The internal vaporisation energy/molar volume ratio E/V_{mol}, is referred to as the **molar cohesive energy density**. This quantity is a measure of the energy input required in the process of dissolution to overcome the cohesive forces in the solvent or solute. Its square root is the concept of the solubility parameter δ (Equation 4.40), which was introduced by *Hildebrand* [94] and later by *Burrell* [95, 96], and which can be subdivided analogously into the quoted energy components δ_D, δ_P and δ_H. We may thus proceed from the following relationships:

$$\delta^2 = \frac{E}{V_{mol}} = \frac{E_D}{V_{mol}} + \frac{E_P}{V_{mol}} + \frac{E_H}{V_{mol}}$$

Equation 4.40:

$$\delta = \sqrt{\frac{E}{V_{mol}}} = \sqrt{\delta_D{}^2 + \delta_P{}^2 + \delta_H{}^2}$$

If the vaporisation energy is expressed in calories and the molar volume in millilitres, the solubility parameter has the unit *Hildebrand* (*h*), 1 *h* being 1 $(cal/ml)^{0.5}$.

Good solvency of a resin in a given solvent is guaranteed if the cohesive forces of the resin are in balance with those of the solvent, and this is most likely to be the case when they have equal or similar solubility parameters. However, similarity or equality of the solubility parameters is an indication, but not assurance of good solubility or compatibility.

Provided that the vapour in question can be considered to be an ideal gas, the change in internal energy during vaporisation can, in many cases, be calculated from the vaporisation enthalpy as ΔE is equal to ΔH - RT = ΔH - $p\Delta V$. ΔV is in this case the difference between the molar volumes of vapour and liquid at the thermodynamic temperature. Further methods for assessing the solubility parameter start from the chemical structure of the substance, or compare its solubility behaviour with that of other liquids with known solubility parameters.

Table 4.10: Examples for the solubility parameter values, h, for several fluids and synthetic resins

Substance	Solubility parameter [h]
White spirit	7.8
Xylene	8.8
Acetone	9.8
n-Butanol	11.3
Methyl ethyl ketone	9.3
Butyl acetate	8.5
Butanol/xylene 1 : 1	9.2
Water	23.5
Nitrocellulose	7.8 to 14.7
PVC/PVA-copolymer	7.8 to 13.8

A survey and description of these methods can be found in the book of *Patton* [78] mentioned previously and in a paper by *Burrell* [95]. Articles by *Sørensen* and *Eissler* et al. [97, 98] discuss application possibilities of the solubility parameter in practice. The values of the solubility parameters for some solvents and binders are listed in Table 4.10.

In the calculation of the solubility parameter, *Brandts-Buys* and *Huyskens* do not consider it correct to treat the hydrogen bonding forces in the same way as the dispersion and polar forces. Whereas the application of the geometric-mean principle can be justified for the latter, this seems not possible in the case of hydrogen bonds, because, due to their nature, the necessary pre-requisites for using that principle are not given [99].

Further aspects of solvent influence

The acid-base concept discussed above is also a good means of studying binder/solvent interactions, because the solubility of resins is similarly governed by their acid-base character.

Partly on account of their own viscosity and partly due to their interactions with resins, solvents also have a considerable influence on the viscosity of the medium. Solvation occurs when the forces between the solvent molecules and the macromolecules are greater than the cohesive forces between the solvent molecules. This causes swelling of the macromolecules and thereby an increase in viscosity. If the solvent present is unsuitable, poor or even no solvation at all occurs and the macromolecules remain more or less in a coiled state in the resin solution, resulting in a lower viscosity.

Theories of colloidal stability by steric hindrance

Interaction energy and particle separation

As two dispersed particles covered with an adsorbed layer of macromolecules approach in the liquid medium, forces between them come into effect that, depending on type and extent, may lead to flocculation or stabilisation. Hence, the particles must possess potential repulsive and attractive energies whose action is strongly determined by their distance of separation. The attractive energy arises from the intermolecular forces. The repulsive energy originates in the adsorbed layer and is the result of the volume restriction in the interpenetration zone of the polymer layers on the one hand, and of the herewith implied osmotic and thermodynamic effects on the other hand. The resulting interaction energy, termed the **total interaction energy**, will be the sum of the repulsive and attractive forces. The scheme in Figure 4.8 illustrates the variation of the total interaction energy between two particles as a function of their distance of separation.

By convention, the potential repulsive energy E_R is defined as being positive, while the potential attractive energy E_A is considered to be negative. E_R decreases exponentially with distance, whereas E_A as a first approximation diminishes inversely proportional to the square of the distance. The total interaction energy is represented by the solid curve in Figure 4.8. Its maximum corresponds to the energy barrier that the particles should overcome in order to collide. Although the total interaction energy is determined over a long range of values by the repulsive energy, the attractive energy (van der Waals attraction) predominates at extremely small distances. Both modes of interaction, E_R and E_A, become weaker as the distance of separation between the particles increases, until at sufficiently large distances there is no further interaction.

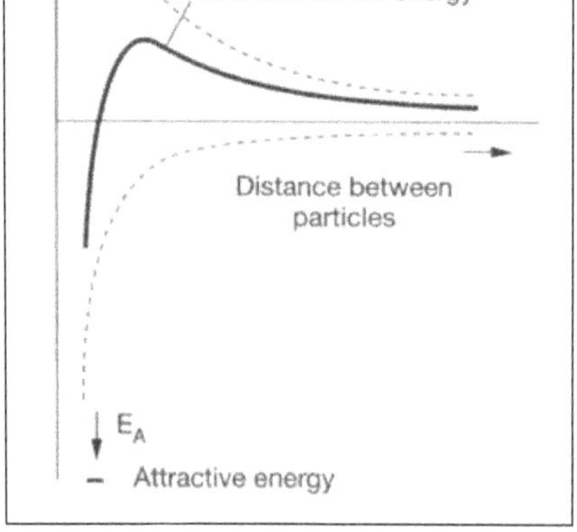

Figure 4.8: Repulsive energy, attractive energy and total energy of interaction in dependence on particle distance (schematic)

Dependence of the attractive energy on distance and particle size

A general accepted explanation of steric hindrance is that the adsorbed polymer layer imposes a merely mechanical barrier between the particles. At sufficient thickness of the layer, the particles are prevented from approaching closely

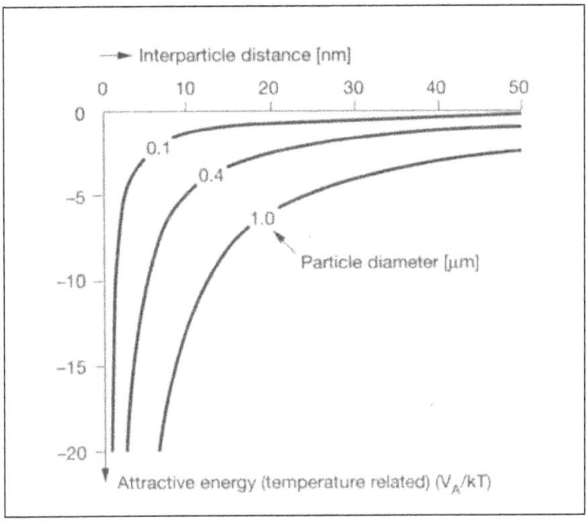

Figure 4.9: Attractive energy according to the Hamaker equation in dependence on particle distance and size

enough to allow the attractive forces to take effect. The Brownian motion of the particles then suffices to separate them again [100].

The attractive forces between two particles are, apart from their type and distance of separation, also dependent on their size. As mentioned above, the distance is determined by the thickness of the adsorbed layers. The variation of attractive energy (in *kT* units) with the distance of separation is plotted in Figure 4.9 for three particle sizes. It should be noted that, according to the kinetic theory of gases which also applies to colloidal systems, the term *kT* is the **most probable kinetic energy** imparted to a particle and the standard unit of thermal energy. *k* is the Boltzmann constant, equal to $1.3807 \cdot 10^{-23}$ Joule/Kelvin and T the thermodynamic temperature in Kelvin. In this context, the question arises as to why the most probable energy is employed in these relationships instead of the average kinetic energy (2/3) *kT*, which would actually be more exact.

It can be seen from Figure 4.9 that at equal separation the attractive energy between the particles with 1 µm diameter is about twice as high as is the case for particles with 0.4 µm diameter. Hence, smaller particles can be stabilised with thinner polymer layers.

The potential attractive energy between dispersed particles can be calculated by means of a relationship stated by *Hamaker*. He assumed that the interaction between individual particles is an additive effect composed of the interaction forces between atoms or molecules, and that it could possibly be calculated by integration over all pairs of atoms or molecules. The energy potential of the attractive forces taking effect between two atoms (molecules) separated by a distance *d* is:

$$E_A = -\frac{\beta}{d^6}$$

In this equation, β is the London constant, which depends on the kind of atoms (molecules). It consists of the number of electrons in the outer shell, Planck's constant h, the characteristic oscillation frequency and the polarisability of the atoms (molecules). *Hamaker* derived the following expression for two spherical particles in vacuum:

$$E_A = - \int_{V_1}\int_{V_2} \frac{q^2 \beta}{d^6} dV_1\, dV_2$$

where

V_1 and V_2 = particle volumes and
q = number of atoms or molecules per millilitre.

The integration of this equation for two spherical particles of equal size with radius R, separated by a distance d in a vacuum gives:

Equation 4.41:

$$E_A = -\frac{A_H}{6}\left[\frac{2}{s^2-4} + \frac{2}{s^2} + \ln\left(\frac{s^2-4}{s^2}\right)\right]$$

Here $s = d/R$, i.e. the ratio of the distance between the particle centres to the particle radius. The expression between the square brackets is thus governed only by the geometrical conditions. A_H ($= \pi^2 q^2 \beta$) is the *Hamaker* constant, given by the number of atoms or molecules per millilitre q and the London constant β. A_H can be calculated from the refraction indices. Its values vary between 10^{-19} and 10^{-21} J. For very short distances between the surfaces of the spheres, i.e. when $d - 2R \ll R$, Equation 4.41 becomes:

$$E_A = - \frac{A_H R}{12(d-2R)}$$

In a pigmented medium, the potential attractive energy between the particles, arising from medium-medium and medium-pigment interactions, must be weaker than in vacuum. Hamaker was able to demonstrate that these interactions only appear in the Hamaker constant in Equation 4.41 and that, therefore, the value of the composite Hamaker constant is given by:

$$A_H = H_{PP} + H_{MM} - 2H_{MP}$$

where H_{PP}, H_{MM} and H_{PM} represent the corresponding interactions between particles, between medium molecules and between particles and medium molecules. Since, according to *London*, the interaction between two different molecules is approximately equal to the geometric mean of the forces involved, i.e. $H_{PM} = \sqrt{H_{PP} \cdot H_{MM}}$, it follows that for the Hamaker constant in a dispersion with homogeneous pigment particles:

$$A_H = (\sqrt{H_{PP}} - \sqrt{H_{MM}})^2$$

Concerning the attractive energy when the polymer layer adsorbed on the particles is taken into consideration, the interested reader is referred to a publication by *Vold*, which contains a detailed discussion of this problem [101].

Stabilisation by entropy reduction

A theory of steric hindrance by entropic repulsion was propounded by *Clayfield* and *Lumb*. They attribute the stabilising effect to the fact that when the adsorbed layers of two approaching particles overlap, the mobility of the free polymer segments is impaired, implying a reduction in entropy [102].

Depletion stabilisation

A further possible explanation of steric hindrance is provided by the theory of **depletion stabilisation**. It postulates that the dispersed particles are prevented from colliding with each other not so much by polymer adsorption, but rather by the presence of free suspended macromolecules. It is assumed that as two particles approach, coils of macromolecules located between them become compressed, involving an impairment of their mobility and hence a decrease in entropy. This is supposed to give rise to repulsive forces, which at sufficiently high polymer concentration are enough to keep the particles apart (Figure 4.10). Initial theoretical research work on this phenomenon carried out by *Feigin* and *Napper* does not as yet allow any definite conclusions as to the exact facts [103].

Thermodynamic interpretation of the steric stabilisation

According to *Napper's* thermodynamic treatment [83], when two particles in a suspension enveloped in polymer layers closely approach, the free enthalpy must be positive in order to obtain repulsion, i.e. so that the process may spontaneously run in the opposite direction. The free enthalpy comprises an enthalpic and an entropic contribution. The relationship between free enthalpy G, enthalpy H and entropy S is, according to *Gibbs* and *Helmholtz*, given by the equation:

Equation 4.42:

$$\Delta G = \Delta H - T\Delta S$$

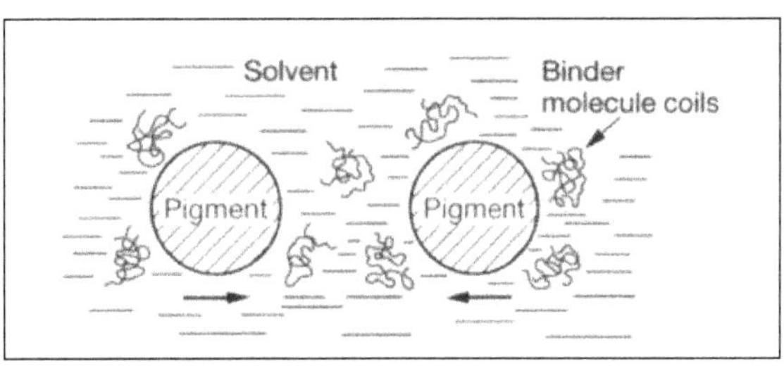

Figure 4.10: Schematic representation of depletion stabilization

A negative change in the free enthalpy ($-\Delta G$) means support for the process of approchment by the entanglement of the unbound polymer

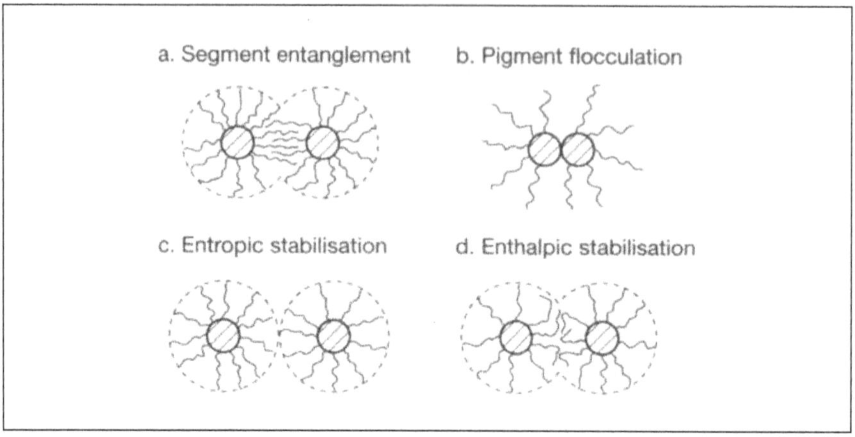

Figure 4.11: Influence of free enthalpy on stabilisation: (a) flocculation by entanglement of unbound polymer segments; (b) free pigment flocculation, neither impaired nor supported; (c) entropic stabilisation; (d) enthalpic stabilization

segments. Spontaneous flocculation follows (Figure 4.11a). When the adsorbed layers do not overlap or entangle, flocculation – neither impaired nor supported – may occur as if the adsorbed layers were not present at all. In this case $\Delta G = 0$ (Figure 4.11b).

When the value of ΔG is positive, flocculation cannot take place; hence, the system is sufficiently stabilised. Depending on the amount and sign of the enthalpic and entropic contributions (Equation 4.42), this state can be attained in three different ways:

1. Both enthalpic and entropic effect are negative, but $T\Delta S$ is greater than ΔH. Consequently, ΔS opposes flocculation while ΔH favours it. This case is termed **entropic stabilisation.** Cooling of the system leads to flocculation, as temperature reductions imply a decrease in the entropic contribution (Figure 4.11c).

2. Enthalpic and entropic effect are both positive, but this time $T\Delta S$ is weaker than ΔH. Since the stabilisation effect here is due to ΔH, such a system is said to display **enthalpic stabilisation**. It flocculates on heating, because with rising temperature the entropic contribution increases (Figure 4.11d).

3. The enthalpic contribution is positive and the entropic one negative, i.e. both support stabilisation. In this case we have a combined **enthalpic/ entropic stabilisation**. Such a system is not subject to flocculation due to changes in temperature over a wide range.

The following Table 4.11 by *Napper* summarises the three ways described above which may lead to a positive ΔG value.

Table 4.11: Relationships between the thermodynamic state functions ΔH and ΔS bringing about steric stabilisation *Source: D.H. Napper [83]*

ΔH	ΔS	ΔH/TΔS	ΔG	Type	Flockulation
-	-	<1	+	entropic	by cooling
+	+	>1	+	enthalpic	by heating
+	-	$\gtreqless 1$	+	enthalpic-entropic	not realizable

A deficiency in this so-called **interpenetration model** is that it does not make provision for mechanical properties and the resulting interaction between the adsorbed polymer layers.

In general terms, experts agree nowadays as to the fundamental principles of steric hindrance; however, its complex mechanisms are, in contrast to electrostatic stabilisation, less accessible to mathematical interpretation and have to date not been completely clarified.

4.2.3.3 Electrostatic stabilisation: DLVO theory and zeta potential

The second stabilising option is based on the electrostatic repulsion between particles with surfaces carrying the same charge. The adsorbed polymer shells are the charge carriers. By the specific adsorption of ions having the same positive or negative charge onto their surface, the particles wrap themselves in a charge layer which may also include water molecules (in the case of waterborne systems). In its turn, this charge layer attracts an oppositely charged cloud of ions. Thus, **electric double layers** result that start to interpenetrate when the particles come close to each other and thereby give rise to repulsion effects. The Coulombic repulsion can, however, take effect in suspensions containing solely water or strongly polar organic solvents.

The theory of electrostatic stabilisation of colloids was developed by *Deryaguin* and *Landau* and independently by *Verwey* and *Overbeek*, and is thus known as **DLVO theory**. It allows calculation of the magnitudes of the attractive and repulsive forces in dependence upon the parameters involved, and is also valid for charged particles in non-polar or weakly polar systems [104, 105].

The formation of electric double layers at the interface between disperse phase and polymer solution was first considered by *Helmholtz*. He interpreted the electric double layer as a molecular plate capacitor. One plate consists of the ions attached to the surface and the other plate is formed by the ions of opposite charge in the dispersion medium. In the case of charged pigment particles, a molecular capacitor composed of two concentric spheres with the same charge density but opposite charge signs, i.e. as shown in Figure 4.12 (above), is likely to give a better representation of the real conditions.

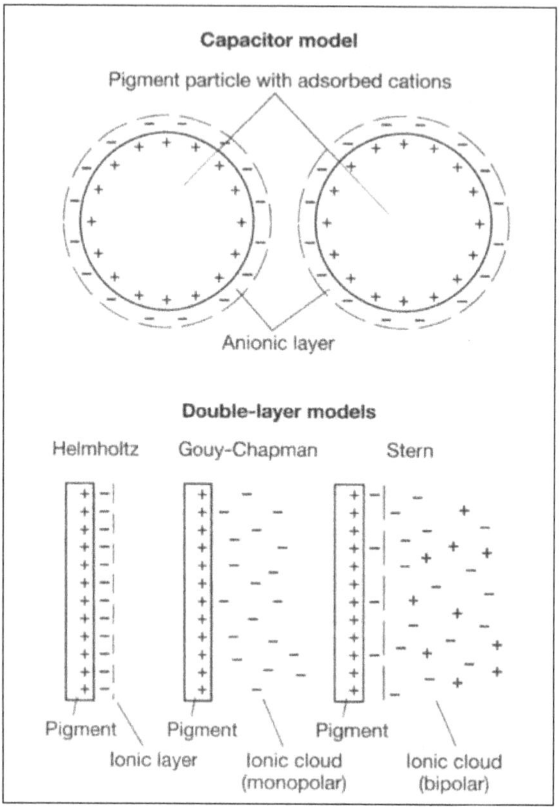

Figure 4.12: Helmholtz interpretation of the electrical double layer on a disperse particle as 'plate capacitor' consisting of pigment particles with cation and anion layers attached to their surfaces (above). Electric double layers according to the interpretations of Helmholtz, Gouy-Chapman and Stern (below)

Measurements of the potential difference according to *Helmholtz's* model showed that this model does not always reflect the true facts. Research by *Gouy* and *Chapman* demonstrated that there is no sharply defined delimitation for the layer of ions in the medium, but that it extends deeply into the polymer solution. This was attributed to molecular Brownian motion, which strives for a uniform distribution of the ions of the outer layer in the dispersion medium. Consequently, the outer layer is not to be considered as a surface, but rather as an ionic cloud. In accordance with this approach, the electric double layer is termed **diffuse double layer**.

Stern was able to prove later that the real situation can best be described by a combination of both models. According to his theory, the ionic cloud actually consists of two parts: The first part is the ionic layer rigidly constrained on the surface. The second part is the diffuse ionic cloud extending deeply into the solution. However, the distribution of ions in this diffuse ionic cloud is not uniform, because the electric field on the particle surface causes an enrichment of oppositely charged ions in its neighbourhood. The three models of *Helmholtz*, *Gouy-Chapman* and *Stern* are represented in a simplified manner in Figure 4.12 (below).

One consequence of the non-uniform charge distribution at the interface is a potential drop between the fixed ionic layer on the particle surface and the medium, i.e. throughout the diffuse ionic cloud. It is assumed that the ionic distribution in the

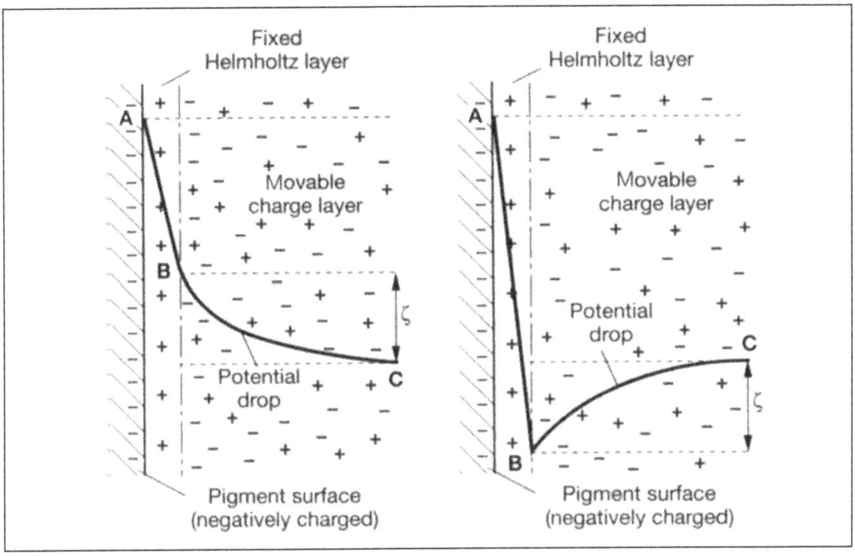

Figure 4.13: Potential drop between particle surface and medium: AB = potential drop in the fixed laver, BC potential drop in the mobile layer.
Left: fixed and mobile layer have the same sign of charge.
Right: fixed and mobile layer have opposite signs of charge.
ζ= Zeta potential

diffuse zone of the double layer can be described by a Boltzmann distribution; i.e. that the potential drop increases exponentially with distance away from the particle surface, as in the *Debye-Hückel* approximation in these authors theory of electrolytic dissociation. The potential drop may arise in two different ways, depending on the specific properties of ions and molecules in the mobile region of the double layer. Both possibilities are shown in Figure 4.13.

In both cases, the potential drop in the fixed ionic layer is given by the straight line AB and the gradual potential decrease in the mobile layer by the curve BC. The latter may appear as in Figure 4.13 left or right, depending on whether the mobile layer has a sign the same as or opposite from that of the fixed ionic layer.

The potential difference accounting for electro-osmosis and other electrokinetic phenomena (e.g. sedimentation and streaming potentials, electrophoresis) corresponds to the potential drop inside the mobile layer and is termed **zeta potential**. It is represented in Figure 4.13 by the sections marked with the Greek letter ζ (zeta). The zeta potential serves to characterise the electrostatic interaction in dispersion colloids, as its value enables inferences as to the surface charge and the potential of the particle surface. It can be assessed, for example, by electrophoretic measurements (see Section 4.2.3.4).

Apart from the ionic charge in the electrochemical double layer, the thickness of this layer is also a determining factor for the electrostatic repulsion and thus for the dispersion stability. High thickness of the double layer occurs when the ionic concentration in the medium is low. By adding an electrolyte, the ionic concentration is raised and thereby the thickness of the double layer is reduced. When the double layer is so thin that the electrostatic repulsion does not suffice to compensate the attractive forces, flocculation of the system results.

In order to calculate the repulsion energy E_R, *Verwey* and *Overbeek* differentiated between the two extreme cases. For systems with coarse particles and a thin double layer, they derived the relationship:

$$E_R = \frac{\varepsilon R \psi_0^2}{2} \ln\left[1 + \exp - \kappa (d - 2R)\right]$$

For systems with small particles and a thick double layer, as well as for media having a low dielectric constant, they gave the simplified version:

$$E_R = \frac{\varepsilon R^2 \psi_0^2}{d} \left[\exp - \kappa (d - 2R)\right]$$

In both equations:

E_R = the repulsion energy
ε = the dielectric constant of the medium
R = the particle radius
ψ_0 = the potential on the particle surface
$1/\kappa$ = the double layer thickness
d = the distance between the centres of two particles.

Both equations, derived for spherical particles, allow an approximate calculation of the repulsion energy. We can infer from them that good stability is obtained when the particles are large and the surface potential is high, as the repulsion energy is directly proportional to the second power of these two parameters. The double layer thickness can be expressed as follows:

$$\frac{1}{\kappa} = \sqrt{\frac{\varepsilon kT}{8\pi n e^2 z^2}}$$

where:

κ = the reciprocal of the thickness of the double layer
k = the Boltzmann constant ($1.3807 \cdot 10^{-23}$ J/K)
T = the thermodynamic temperature in K

n = the ionic concentration in number per ml
e = the elementary electrical charge (1,60218 · 10^{-19} C)
z = the valency of the ions.

In many cases, particularly with non-aqueous systems, the surface potential can be replaced by the zeta potential. One shortcoming of the DLVO theory is that these equations do not take into consideration the interaction of the ions with the medium. The reliability of the pertinent assertions and thus their application are also impaired by the fact that at high concentration two particles may interact not only with each other, but also with particles in their immediate neighbourhood. This is also disregarded in the corresponding equations. Despite these imperfections, the DLVO theory allows us to gain an insight into the relationships of electrostatic stabilisation and the means for their evaluation. This is only approximate, it is true, but in many cases sufficient for practice.

4.2.3.4 Measuring Zeta potentials

When suspensions of electrostatically charged particles are placed into an electric field of field strength E, they will migrate to the oppositely charged electrode with the velocity n. The quotient of n/E is called the electrophoretic mobility μ. The electrophoretic mobility is connected to the zeta potential ζ via the viscosity η of the liquid and its dielectric constant ε. Depending upon the relationship between the thickness of the double layer and the particle size, two cases may be distinguished. When the extension of the double layer is small in comparison to the size of the particle diameter:

$$\zeta = \frac{\mu}{\eta \cdot \varepsilon}$$

This is normally the case in aqueous media (*Smoluchowski* approximation). When the double layer is thicker than the particle diameter, the relationship is extended by a factor 2/3 (*Hückel* approximation).

$$\zeta = \frac{2\mu}{3\eta \cdot \varepsilon}$$

Traditionally, zeta potentials were determined using microelectrophoresis. The electrophoretic mobilities of the particles were measured under a microscope while an electrical field was applied. These measurements were tedious and had certain limitations. Dispersed pigment particles, for example, are mostly submicroscopic so that it is not possible to observe primary particles under the microscope. In addition, it was only possible to measure in very dilute suspensions.

An improved method to determine the migration of the particles is laser Doppler anemometry. In this technique, a laser light beam is split and then brought together in a crossing point within a suspension again. By interference, a grid of light and dark stripes is generated. Once again, an electric field is applied causing charged particles to migrate. Every time they move through light areas, they scatter light. Their speed is then determined from the frequency of the light signals.

Today, electro-acoustic methods are widely used. Three methods rely either on:

- applying an a.c.-electrical field to a suspension of charged particles and measuring pressure waves that are generated (*ESA*, electrokinetic sonic amplitude), or
- applying ultrasonic waves to the suspension and measuring the resulting a.c. potential *UVP*, ultrasonic vibration potential or
- CVP, colloidal vibration potential.

In *ESA* measurements, the alternating electrical field causes the particles to move back and forth. In doing so, they generate pressure waves. One of the electrodes is mounted on a piezoceramic. The amplitude of the resulting pressure wave P divided by the electrical field strength E is proportional to the high frequency electrophoretic mobility μ_D, the density difference between the particles and the liquid $\Delta\rho$, the volume fraction of particles ϕ and the speed of sound in the liquid c [182].

$$ESA = \frac{P}{E} = \Delta\rho \cdot \phi \cdot c \cdot \mu_D$$

Since all parameters are either known or can be independently measured, the high frequency electrophoretic mobility can be determined. From that, the zeta potential is found via Equation 4.43 [183]:

Equation 4.43:

$$\zeta = \frac{\mu_D \cdot 3\eta}{\varepsilon} \cdot \frac{1}{G_\alpha}$$

G_α is an inertia term. It is found from the lag between the electric field strength and the pressure wave. G_α is related to the mean particle size. Larger particles have a higher inertia which results in a larger time lag between the two. It is normally found that the particle size passes through a maximum at the isoelectric point where the particles flocculate. Although *ESA* measurements are more or less independent of the concentration of particles in a suspension, particle size measurements may very well depend upon the volume fraction of particles since the inertia of the particles will increase if the particles are concentrated highly enough to interact with each other. While measuring cement suspensions, it was found that the particle sizes measured increased from 7 μm to 27 μm when the volume fraction was raised from 35 % to 42.5 % [183].

In ultrasonic vibration potential (*UVP*) measurements, ultrasonic sound waves are used to cause the particles to oscillate in the suspensions. Even more so, their electrical double layers, which have far less inertia, follow the pressure waves so that the electrical double layer is periodically distorted relative to the particles in the frequency of the pressure wave, thus causing a potential ψ within the suspension. The *UVP* or CVP (colloid vibration potential) is reported in the unit volts per velocity amplitude U_0 of the applied acoustic field. It has the physical dimension V/(m/s). For a parallel plate geometry, the *UVP* and its relation to the *ESA* signal is given by

Equation 4.44:

$$UVP = \frac{\Psi}{U_O} = \frac{\Delta\rho \cdot \phi \cdot c \cdot \mu_D}{K^*} = \frac{ESA}{K^*}$$

Equation 4.44 shows that the high frequency conductivity of the suspension K^* connects the *ESA* and the *UVP* signal. An advantage of the *ESA* measurement in comparison to the *UVP* measurement is that K^* is not needed to calculate the electrophoretic mobility. Especially in media with a low dielectric constant, K^* is difficult to determine accurately.

4.3 Stabilisation problems with pigmented systems after dispersion

Certain negative effects occurring only during the later stage of processing, e.g. the application or storage of pigment dispersions, are actually connected with the stabilisation of the system. However, due to their importance they will be discussed in this separate section which refers to the phenomena termed **flocculation, flooding, floating, shock effects and seeding**.

4.3.1 Flocculation

4.3.1.1 Fundamental aspects

The effect known as flocculation was mentioned briefly in the preceding Section 4.2.3. It is ascribed to inadequate stabilisation that is insufficient to shield the attractive forces at the particle surface. The pigment particles in the system,

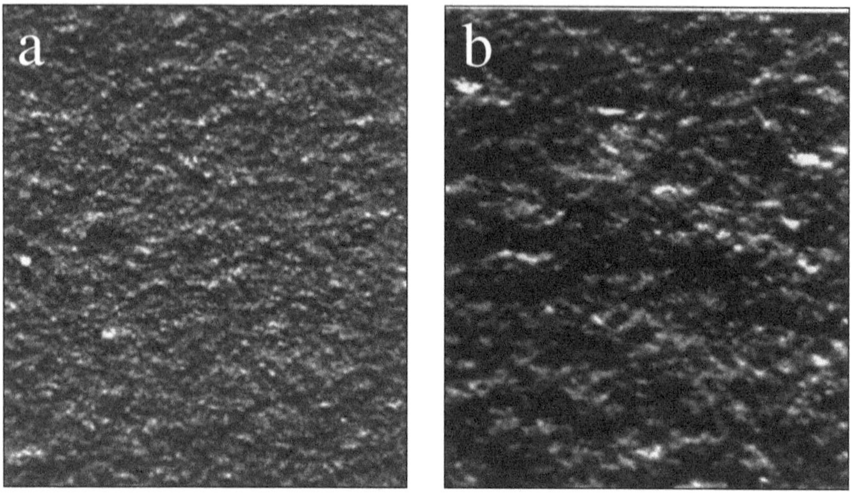

Figure 4.14: Scanning electron micrograph (x1000) of a coating surface containing C.I. Pigment Red 3: a) standard grade, b) flocculating batch

Source: Works micrograph Hoechst AG

apparently well dispersed initially but de facto partially wetted and poorly or weakly wrapped, form loosely bound clusters called **flocculates** or **flocs**. This generally unwelcome process gives rise to an increase in yield value, thixotropy and rate of sedimentation in the fluid system on the one hand, and to variations in shade and a loss in depth of shade, gloss, semi-transparency or hiding power in the dried film on the other hand.

Hence, flocculation partially or completely reverses the result of the dispersion operation. Moreover, the pigment flocculates have the unpleasant property of being readily broken down by the action of weak mechanical forces, so that the optical properties become labile and unpredictable. In multi-pigmented systems, flocculation results in pigment separation effects known as flooding and floating. These phenomena can, however, also have other causes apart from flocculation, as will be discussed in the next section. All these reasons lead to flocculation being almost always an undesired phenomenon for the pigment processor.

A conception of what is understood by flocculation is conveyed by the scanning electron micrographs in Figure 4.14. They show dry paint films containing an organic pigment (C.I. Pigment Red 3). The standard grade is shown in 4.14a and a flocculated production batch departing from standard in Figure 4.14b. The flocculation in the latter case has led to a rugged, practically dull surface.

Due to their smaller particle size and thus larger specific surface area, organic pigments have a stronger tendency to flocculate than most inorganic pigments; among the latter, products with high fineness of particle, e.g. Prussian Blue and many carbon-black grades, are particularly prone to clustering again. This is probably because they have a lower surface polarity and therefore undergo less interaction with polymeric components of the formulations. However, the pigment itself is not solely responsible for this phenomenon. The binder/solvent system, the dispersing conditions and the presence of other pigments also influence flocculation. One and the same pigment in distinct systems may thus show differing flocculation behaviour. Due to the variety of influencing factors, this paint defect, as a rule, cannot be forecast and must be investigated after the dispersion operation has been completed.

Nevertheless, it would be wrong to consider flocculation to be a phenomenon only resulting in disadvantages and to be eliminated at all costs. Certainly, as a quantitative phenomenon, flocculation, when occurring to a large degree, is detrimental to all the above-mentioned features of the pigmented system. However, a slight degree of flocculation may be desirable in many cases, as it unites the advantages of good stabilisation without shortcomings such as the formation of a hard sediment and poor sagging behaviour [106].

4.3.1.2 Testing and measuring flocculation behaviour

It should be noted that, when investigating flocculation behaviour, it is advisable to assess the flocculation rate from several flocculation-dependent properties rather than from only one.

Sedimentation test

One method for judging flocculation behaviour is based on the fact that flocculates behave similarly to agglomerates, and settle out more rapidly than aggregates and primary particles. This is explained by *Stokes'* law, according to which the degree of settlement is directly proportional to the square of the particle diameter. Consequently, the larger and more numerous the flocculates, the higher the rate of settling.

The pigment dispersion is diluted in order to carry out the test, as Stokes' law is only applicable to diluted systems. Dilution also reduces the viscosity and thereby accelerates the sedimentation. The diluted paint is poured into a graduated measuring cylinder and stored for a period of time. A pigment-free layer gradually develops with flocculated systems, and this layer generally exhibits a clear boundary so that its height can be easily determined from the scale. In contrast, stable, colloidal systems show a transition zone between the clear, supernatant binder solution and the pigmented phase. The nature of the sediment can be probed by stirring with a glass rod or steel spatula (as specified in ASTM D 869). Flocculated systems form fluffy and voluminous sediments by virtue of their low density of packing, whereas the deposits of properly stabilised dispersions are normally compact and cannot be easily re-incorporated into the system by stirring.

Although this method gives fairly clear results, dilution represents a critical intervention into the system and often leads to the question of whether the system would behave exactly the same in an undiluted condition.

A further means to accelerate the sedimentation process is by keeping the dispersion for a couple of hours in warm storage, as the higher temperature raises the mobility of the particles. Acceleration of settling by centrifuging techniques generally affects the consistency and volume of the sediment and thus should not be used.

Pour test

According to another method, the dilute pigmented system is poured onto a glass plate or transparent foil. After drying gloss and transparency are assessed, both being properties that deteriorate considerably with flocculation. Comparison of poured specimens prepared before and after storage allow conclusions to be drawn as to the degree of flocculation. Pigments resistant to flocculation show very little or no difference. Due to the dilution, this method is subject to the same limitations as the sedimentation test and is only reliable if one can be confident that, apart from the lower viscosity, no further drastic changes in the dispersion have occurred. In such cases, the best approach is dilution with the system binder or binder solution.

Rub-out test

A further, relatively simple method of assessing flocculation is the so-called **rub-out test**. It consists of applying the fluid pigmented medium (e.g. by spraying, brushing or drawing down) to a suitable substrate; for example, a steel panel. After the main part of the solvent has been allowed to flash off, a portion of the wet coating is heavily rubbed with a brush or the finger. The mechanical forces thus exerted are sufficient to break down any flocculates formed, and the rubbed surface will exhibit a change in shade. In the case of coloured pigments, the shade will appear deeper if flocculation has taken place. If the tested pigment is white, the rubbed area will show a higher lightness as the light is more strongly scattered. This method is mainly applicable to inorganic pigments, such as red and yellow iron oxides and chrome oxide green. The high tinting strength of organic pigments leads to variations in the depth of shade caused by flocculation to be hardly noticeable in mass tones. More on the rub-out test can be found in Section 4.3.2 in connection with flooding and floating.

Chromatographic method

A chromatographic method for determining flocculation has been proposed by *Kresse*, and is based on the fact that monodisperse particles, in contrast to flocculates, migrate together with the binder solution. A strip of filter paper dipped into the system being tested thus shows a pigment-free zone at the top if flocculation has occurred [107].

Assessment of the degree of flocculation

The quantitative assessment of flocculation can be adequately performed with the aid of other methods such as measurement of gloss reduction, variations in the scattering power or of scattered infrared radiation. If the particle diameter is higher than the paint-film thickness, the particles emerge from the coating surface. This gives rise to scattering centres that impair the specular reflection and cause a decrease in gloss. In other words, flocculates behave exactly the same as agglomerates in the dried paint film, and it may easily be seen that this kind of gloss reduction is dependent on the film thickness. Hence, flocculation has a stronger effect on the gloss of thin paint films than in the case of thick layers. More details about gloss and gloss measurement are given in Section 2.1.

The principle of the second method is based on a decrease in the scattering coefficient resulting from the flocculation of all those pigments that cause light scattering in the medium, i.e. white and most of the inorganic coloured pigments. Organic pigments are in general manufactured with a view to high tinting strength and low light scattering, and thus in their case flocculation results in an increase in scattering power. A suitable method, founded on the dependence of scattering power upon the degree of flocculation, has been developed by *Keifer* [108].

The last method quoted for determining the degree of flocculation is based on the fact that coloured pigments exposed to infrared radiation behave like white pigments in the visible spectrum, i.e. infrared radiation is not absorbed but merely scattered by coloured pigments. At a given film thickness, PVC and wavelength, IR-scattering increases with the number and size of the flocculates. *Balfour* and *Hird* described this method and demonstrated its possibilities by means of experimental examples [109].

4.3.1.3 Kinetic approach to flocculation

Assuming that each collision between two particles produces a flocculate, the reduction in particle concentration with time, i.e. the flocculation rate, is dependent upon the particle concentration. It is thus possible to consider the flocculation process as a second-order reaction. If we term the number of particles per unit volume N, the rate of flocculation is given by the following equation:

$$-\frac{dN}{dt} = k \cdot N^2$$

The decrease in concentration with time is thus proportional to the second power of the concentration. The constant k has its maximum value when no repulsive forces occur to inhibit the flocculation process. The following relationship was determined by *von Smoluchowski* for the extremely idealised case of spherical, monodisperse particles in a diluted dispersion where no repulsive forces may take effect and each collision gives rise to a flocculate:

Equation 4.45:

$$-\frac{dN}{dt} = 8\pi RDN^2$$

in which R is the collision radius and D the diffusion coefficient [110]. For uniform particle size, the collision radius is equal to twice the particle radius. In the case of spherical particles, the diffusion coefficient is given by the equation of *Einstein* and *von Smoluchowski*:

Equation 4.46:

$$D = \frac{kT}{6\pi\eta r}$$

where k is the Boltzmann constant, T the thermodynamic temperature, η the viscosity of the fluid phase and r the particle radius. The term in the denominator is the frictional index in viscous media defined in Stokes' law (see Section 1.3.1.4). If it is considered that each collision of two original aggregates results in a flocculate that is also capable of colliding with a further aggregate or flocculate, it is easy to understand that the decrease in concentration of all particles existing in the system at the beginning is actually only half that given by Equation 4.45:

Equation 4.47:

$$-\frac{dN}{dt} = 4\pi RDN^2$$

Hence, for the period in which flocculation takes place:

$$t = \int_{N_0}^{N} \frac{1}{4\pi RDN^2} dN$$

By integration and introduction of the condition $N = 0.5\ N_0$ in the equation (N_0 is the initial concentration of particles at time $t = 0$), we obtain the time $t_{0.5}$ for the half value, i.e. the time required for the initial particle concentration to be reduced to one-half by flocculation:

Equation 4.48:

$$t_{0.5} = \frac{1}{4\pi RDN_0}$$

Now, taking the collision radius to be twice the radius of the particles and introducing the value of D given by Equation 4.46, the time required for the half value will be:

Equation 4.49:

$$t_{0.5} = \frac{3\eta}{4kT N_0}$$

The Smoluchowski theory is valid for rapid flocculation or coagulation of monodisperse systems. Equation 4.49 tells us that flocculation is favoured by high temperature and high particle concentration, and is impaired by high viscosity of the fluid phase. In polydisperse systems flocculation takes place more rapidly than in analogous monodisperse systems, because the probability of collision increases with the widening range of particle size.

However, one of the essential pre-requisites of the Smoluchowski theory is not correct; namely, that each collision of two particles must result in the formation of a flocculate. It should rather be assumed that two particles coalesce into a larger conglomerate only when their potential attractive energy is sufficient to overcome energy barriers possibly present in the system (e.g. viscosity).

4.3.1.4 Coflocculation

Coflocculation takes place in multi-pigmented systems when two or more pigments flocculate at the same time and at more or less equal flocculation rates, i.e. flocculates with a structure formed by at least two pigments result. The interaction of specific attractive forces between the pigments is not a pre-condition for the achievement of this phenomenon. A previous hypothesis attributing the phenomenon to opposing electric charges of the disperse particles could not be verified in practice. Daniel assumes that conveniently placed sites of opposite polarity are responsible for coflocculation. These result from the heterogeneous character of the pigment surface [111].

Strong coflocculation results in the same disadvantages as strong flocculation, but a slight coflocculation may be just as worthwhile for the pigment processor as slight flocculation. Separation processes can be thus avoided or at least impaired, so that the homogeneity of distribution remains intact. In addition, when the pigment combination contains a white pigment, the usual case with enamels, its particles assume the role of flocculation nuclei onto which the smaller coloured pigment particles accumulate. This leads to better utilisation of the coloured pigment's absorption and to a reduction of the frequently noticed loss of tinting strength due to flocculation. However, one detrimental consequence of coflocculation processes is the decrease in hiding power of the white pigment. The pigment user must thus carefully consider the pros and cons of each particular case, and find a compromise.

4.3.1.5 Prevention of flocculation

As a basic point, the use of flocculation-proof pigments rather than non-treated products can be expected to lead to better results with regard to dispersion stability [112]. A good example of this is provided by the various phthalocyanine types which were rendered stable against flocculation by partial chemical modification (sulphonation, sulphamidation) or by milling the raw product with salt in the presence of xylene or related hydrocarbons [113]. The introduction of polar groups in the phthalocyanine molecule, thus providing a charge on the particle surface, results in mutual repulsion that might oppose the van der Waals' attraction forces.

Section 4.2.3 is related to the main factors that promote the stabilisation of pigment dispersions and thus obviate flocculation. Further measures which can be taken for waterborne systems are the addition of either brine or polyelectrolyte solutions, influencing the surface charge and adjusting the acid-base character of the pigment surface to the medium. The latter possibility can also be applied to non-aqueous media [27, 91, 114, 115].

Finally, the use of surfactants as a means for impairing or eliminating flocculation needs to be mentioned. This possibility will be discussed below (see Section 4.3.2.3).

4.3.2 Flooding and floating

4.3.2.1 Discussion and theoretical explanation

Definitions

The terms flooding and floating refer to processes of separation occurring between different pigments in the wet paint film after application or in the container during storage. This separation can give rise to uniform changes of colour and depth of shade due to enrichment of the film surface by one component of the pigment mixture, or to the formation of surface structures (cells or streaks). In the first case we speak of **vertical flooding** or just **flooding**; in the second case of **horizontal floating** or just **floating**. The expressions **primary** and **secondary flotation**, or just **flotation** meaning floating are occasionally found in the technical literature. For the sake of simplicity and in order to avoid possible confusion, it is advisable to follow *Patton's* approach and use solely the terms flooding and floating, sufficiently specified in the standard DIN EN ISO 4618. Other designations should not be used. Whether pigment separation takes effect in a system as flooding or as floating depends on which of the numerous factors influencing separation (composition of the medium; drying conditions like temperature and rate of evaporation; film thickness, etc.) gets the upper hand.

The effect of such changes in shade and depth of shade is particularly troublesome when two parts of a machine or a car, coated by different application techniques, are assembled next to each other, e.g. when during accident repairs a sprayed panel is mounted alongside an engine cover originally coated by dipping.

Causes

Flooding and floating are caused by differential mobility of the pigment particles resulting mainly from large differences in particle size and in the state of flocculation. Widely differing densities may also play a minor role, but differences in particle size are usually larger than differences in density. A further factor can be electric charges on the surface that are too dissimilar. The opinion is occasionally given in practice that flocculation and the phenomena known as flooding and floating are to be considered as inseparable cause and effect. This is not exact, as the occurrence of such pigment separations cannot always be attributed to flocculation.

Depending on system conditions, an overlap between several separation processes having different, mutually independent causes is also possible.

Separation mechanism

Solvent evaporation at the surface of a wet paint film results in an increase in density, surface tension and viscosity, and in a reduction in temperature. The upper layer of the film sinks downwards, mainly due to differences in surface tension but also assisted by higher weight. The deeper layers, still rich in solvent and of lower surface tension, are displaced upwards and reach the surface. There they spread and their solvents start evaporating. This process, which may be compared to convection in a warmed fluid, recurs over and over again so that a steady flow from top to bottom and vice versa takes place as the wet coating strives to reach a state of equilibrium between the lower and the upper layers. The driving force is predominantly the surface tension gradient between both zones.

When such vortex currents spread over the upper layer of the coating, encounter adjacent flows of the same kind and sink again, the surface takes on a honeycombed structure which consists of five- to six-sided cells, the so-called ***Bénard* cells**. These vortex currents and their effects always occur during such evaporation processes, i.e. also in clear varnishes and particularly strongly in media of low viscosity. In multi-pigmented systems they may lead to separation processes as a consequence of the differing mobility of the pigment particles. In such cases the Bénard cells can be seen with the naked eye. If the painted object stands vertically,

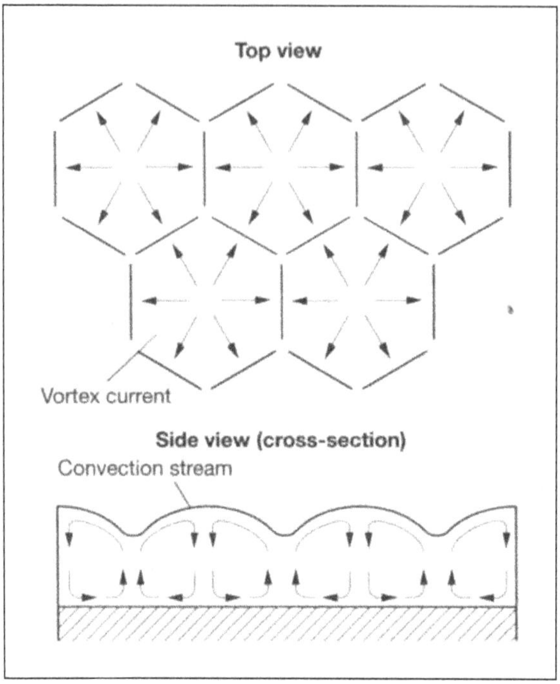

Figure 4.15: Schematic representation of the mechanism of formation of Bénard cells. Top view: Propagation of the vortex flow on the wet layer surface and collision with neighbouring streams. Side view: Convection-like streaming in the coating layer due to differing densities, viscosities and surface tensions in the deeper zones and the surface

Source: T.C. Patton [117]

light and dark parallel streaks are formed instead of the honeycombed structure. The mechanism of Bénard cell formation is schematically illustrated in Figure 4.15. Coatings in which flooding and floating occur are shown in Figure 4.16a and 4.16b (see coloured reproductions on page 217).

Separation processes caused by vortex currents or sedimentation can also be influenced by the pigment volume concentration or the film thickness. At higher PVC, particles mutually impair each other in their mobility. It goes without saying that thicker wet films lead to a more pronounced pigment separation.

Influence factors density and particle size

The greater influence of particle size as compared to density can be demonstrated by taking separation due to differing sedimentation rates as an example. The starting point of our approach is *Stokes'* law, mentioned in Section 1.3 in connection with the settling rate of spherical particles. On studying the settling rates of two pigments in one and the same system, viscosity can be considered to be constant and can therefore be united with the other constants in Stokes' relationship to a unique constant K. Thus, we may express Equation 1.1) as follows:

$$v = K\,r^2(\rho_1 - \rho_2)$$

where

v = the settling velocity
K = a constant
r = particle radius
ρ_1 = density of the pigment
ρ_2 = density of the binder solution

It can easily be seen that the particle radius has a greater influence on the settling rate than the density, as it is present as a squared term whereas the density occurs only as the first power. This is also confirmed by the fact that the heavier pigment occasionally gathers on the surface, whereas the lighter one sinks. To determine whether the difference between densities is responsible for the pigment separation, it is sufficient to allow the paint film to dry with the surface upwards and a second time with the surface downwards. If gravity is the cause of separation, the heavier pigment will gather at the surface in the second case. This has been experimentally proven in practice [116].

In colloidal systems, the mobility B of the particles according to Einstein's and von Smoluchowski's Equation 4.46 is given by the relationship:

$$B = \frac{1}{6\pi\eta r}$$

where h is the viscosity of the medium and r the particle radius. If, for example, the system under consideration is a combination of titanium dioxide with r = 0.150 μm and copper phthalocyanine blue with r = 0.030 μm in a medium with a viscosity of 150 mPa×s, and we neglect the densities (4.1 and 1.6 g/ml) which only play a minor role, we can state that the blue pigment moves with five times the velocity of the white pigment. However, when the blue pigment has coalesced into flocculates with a mean radius of 0.3 μm, the same white pigment particles are twice as fast as the blue flocculates. It is thus understandable that the vortex currents originating in the paint film in the course of flashing-off and drying, or in the container during storage, will always lead to separation processes in a pigment mixture if no other counteracting factors are present.

Summarising the essential points on flooding and floating discussed so far, the following causes may be listed:

1. Differences in particle mobility due to flocculation and/or too great a contrast between the particle sizes.
2. Differing densities of the participating pigments, which may give rise to separation by settling.
3. Vortex currents in the paint layer during drying.
4. Differences in the electrical surface charge of the pigments.

Factors additionally favouring pigment separation are:

5. Low viscosity of the system.
6. Long drying time or storage.
7. Low pigment concentration.
8. High thickness of the wet paint layer.
9. Increased temperature.
10. Overlapping of different types of separation.

It is possible to restrain separation processes by controlling some of these factors, particularly 5, 6, 8 and 9. Other measures will be discussed in Section 4.3.2.3.

4.3.2.2 Testing separation phenomena. spray-pouring test and rub-out test

Spray-pouring test

The main methods for detecting and also for the quantitative estimation of pigment separation are the **spray-pouring** and the **rub-out tests**. The former is based on the fact that different application techniques give rise to different shearing forces taking effect on the paint, or, in other words, the influence of shearing forces on the paint depends on the application method. A coloured enamel is prepared with the test pigment and mixed with a white enamel. The resulting tinted enamel is thinned to spraying viscosity and applied on a suitable substrate, e.g. a card or steel panel. After a suitable flash-off time, the rest of the tinted enamel, after thorough manual stirring, is poured over the lower half of the sprayed coating, allowed to flash-off again and then dried. Differences in hue or depth of shade between the sprayed and the poured area show that pigment separation has occurred.

Rub-out test

In this test, previously mentioned in the context of testing for flocculation (see Section 4.3.1.2), the tinted enamel, prepared as described above, is applied onto a substrate by pouring, spraying or dipping. After a defined flash-off period, which depends on the system, the lower portion of the panel is rubbed with a brush or with a finger until the coating becomes sticky. Care must be taken to ensure that the substrate does not appear through the rubbed surface, i.e. the substrate must remain uniformly hidden. This treatment eliminates any possible pigment separations. The difference between the rubbed and the unrubbed zones is a measure for the degree of flooding.

Both methods can be carried out with air-drying as well as with stoving enamels. The spray-pouring test is more reproducible than the rub-out test, as the latter is more strongly influenced by the tester.

Quantitative evaluation of flooding

In a case of pure flooding, i.e. when no Bénard cells or streaks occur, a quantitative estimation of colour difference is possible by resorting to photometric measurement. The result can be expressed either by *Patton's* **flocculation number** [117] or by the **rub-out number**, as propounded by *Kaluza* [118]. The flocculation number FN is:

Equation 4.50:

$$FN = 100 \frac{(K/S)_A}{(K/S)_B}$$

where $(K/S)_A$ and $(K/S)_B$ are the values for the depths of shade of both surfaces of the test coating, i.e. the unrubbed and the rubbed areas, calculated from the reflectances at the absorption maximum using Equation 2.20, the Kubelka-Munk relationship. As a general rule, $(K/S)_A$ is the depth of shade of the coating surface that was submitted to the weaker shearing forces, i.e. the poured surface in the case of the spray-pouring test and the unrubbed area in the rub-out test.

The rub-out number RN is expressed by Equation 4.51:

Equation 4.51:

$$RN = \left[\frac{(K/S)_A}{(K/S)_B} - 1\right] 100$$

Kaluza speaks of a positive rub-out effect when the rubbed surface is deeper in shade than the unrubbed surface. The alternative is a negative rub-out effect. Using his relationship, a positive rub-out effect results in a positive rub-out number and vice-versa. Tolerances of up to 5 % are usual.

In daily practice, one often hears the opinion that the depth of shade of the rubbed area in the rub-out test corresponds to the depth of shade of the non-flooding or non-floating coating. This assumption is, however, not correct as, on the one hand, the depth of shade depends on the pressure exerted during rubbing and, on the other hand, a pigment separation may even occur in the sticky paint layer – as practice has shown.

4.3.2.3 Methods for overcoming flooding and floating

It follows from the above that basically the mobility of the pigment particles must be reduced in order to prevent pigment separation. On the other hand, possible flocculation of one of the components in a pigment mixture has to be eliminated.

Prevention by adequate selection of raw materials

To minimise flow processes, the differences between densities as well as surface tensions of binders and solvents should be kept as small as possible. Care must also be taken with mixtures of proper solvents with thinners, e.g. in dilutions, to ensure that the evaporation numbers of the components do not differ too widely. The proper solvent should have the higher evaporation number, otherwise resin precipitation during the course of drying may give rise to problems. Moreover, if a system contains several resins, these must be absolutely compatible.

Use of thickening agents

A further measure to reduce pigment mobility is the addition of thickeners. These additives unite the polymer layers adsorbed on the particles' surface with the surrounding binder molecules, thus building up a thixotropic structure that more or less fixes the pigments in the fluid phase. However, this method cannot always be used, as some methods of application, e.g. dipping, do not allow a viscosity increase of the paint.

Use of pigment blends

In particular cases, pigment separation can be prevented by using pigment blends. The intimate mixture obtained by edge-runner grinding of these products opposes the separation of their components. As an example, we may quote **chrome green** (chrome yellow and Prussian blue) and **chrome fast green** (chrome yellow and phthalocyanine blue). Using these mixtures, the pigment processor attains better colour stability than by co-dispersing the relevant yellow and blue pigments.

Employment of surface-active agents

When the pigment separation is caused by flocculation of one of the pigments, it is possible to solve this problem by resorting to surfactants. Due to their chemical structure, these substances gather at the interfaces where they reduce surface tension and thus promote wetting. In addition, surfactants may impair or eliminate pigment flocculation in organic media by steric hindrance and in aqueous systems by influencing the surface charge, i.e. by electrostatic stabilisation, or also by steric hindrance. There are appropriate surfactants for both kinds of systems.

The molecules of a wetting agent consist of a hydrophilic and a hydrophobic group. From the chemical point of view, they may be classified in:

anionic: the hydrophilic group bears a negative charge, as for example alkyl benzene sulphonates, fatty alcohol sulphonates, alkyd aryl sulphonates, as well as the sodium and potassium salts of long-chain fatty acids.

cationic: their hydrophilic group bears a positive charge. In most cases they are salts of long-chain fatty amines.

non-ionic: they have no charge on their hydrophilic group. The polar portion usually consists of multiple oxyethylene groups which are linked to a hydrophobic organic group by a carboxylic or an ether residue. By virtue of their steric stabilising effect they are suitable for preventing settling or flocculation in pigment dispersions.

amphoteric: they can act as either anionic or cationic surfactants, depending upon the pH of the solution. Their application possibilities in pigmented systems are very limited.

With regard to pigment stabilisation in aqueous systems, good results have been obtained with polymeric wetting agents. Thus, sodium polyphosphates, polysilicates and polyacrylates, as well as derivatives of polyvinyl alcohol are recommended for polar pigments, e.g. titanium dioxide. Alkyd-aryl polyglycolic ether derivatives and salts of polymeric aromatic acids such as sodium lignosulphonates have proved useful for organic pigments. The excellent stabilising effect is attributed to the stronger adsorption of these polymeric substances [119].

The best results with solvent-based systems are generally obtained with cationic and non-ionic wetting agents, as for example dodecyltrimethylammonium chloride, soaps of alkanolamines with fatty acids or triethanolamine mono-oleate. Lecithin as well as zinc, potassium and lead naphthenates can also be used for these systems.

Due to the many factors on which flooding and floating are dependent, and in view of the multiplicity of surface-active agents on the market, it is not feasible to give specific guiding rules for their employment. Determination of which surfactants (in this connection termed **anti-flooding** or **anti-floating agents**) are best suitable for a particular system is best carried out by systematic testing in practice. A list of surfactants with their manufacturers can be consulted in *Karstens' Lackrohstofftabellen (Karstens' Tables of Raw Materials for Paints)* [120].

Use of silicone oils

Silicone oils are occasionally used to eliminate floating. Due to their incompatibility with the paint system, these substances form a monomolecular layer on the surface, thereby reducing surface tension and consequently obviating the formation of Bénard cells. However, they prevent neither pigment flocculation nor pigment separation, and wet paint coatings may still show a rub-out effect even after elimination of Bénard cells by the addition of silicone oil. *Kresse* thus recommends the use of silicone oils in combination with other adequate wetting agents in order to obtain satisfactory results [121].

Prevention by coflocculation

A further measure to hinder pigment separation consists of a controlled coflocculation. Special surfactants that promote coflocculation (so-called **anti-floating**

agents) are on the market for this purpose. Because of their very specific effect, their employment demands careful selection and exact dosage.

A study by *Winkler* deals with the stabilising effect of finely-divided fillers, taking blanc fixe as an example. The stabilising mechanism in this case is based on a controlled coflocculation with the pigment to be stabilised. This coflocculation gives rise to a steric shielding of the coloured particles and thus to a reduction in the tendency to separate on flocculation. Coulombic forces may also contribute to the stabilising effect in organic systems [122].

Further information on flooding and floating can be found in relevant articles [100, 123, 124, 125].

4.3.3 Shock effects and seeding

In the course of production of paints and printing inks, the dispersed mill base is generally made up to the final delivery, storage or application composition with further binder, binder solution, additives or mixtures of solvents and diluents. During this stage, termed **let-down**, negative changes may occur in the system if the process is performed without due care. These negative changes, generally known as **shock effects**, can always occur if media with greatly dissimilar concentrations, such as mill base and the let-down varnish, are brought together abruptly. A typical manifestation is the appearance of nibs in the system. Since this phenomenon is generally irreversible, the resulting product is irrevocably ruined.

Depending on the kind of nib formation, a distinction is made between **pigment**, **binder** and **solvent** shock. It should however be pointed out that these denominations are not logical, because in the first two cases the process is named after the affected component, and in the latter case, in contrast, after the originator.

4.3.3.1 Pigment shock

This kind of shock occurs when a pigment-rich and polymer-poor mill base is let down with a binder-rich varnish and results in a renewed agglomeration of pigment particles. Rapid diffusion, mainly of the more mobile component, i.e. the solvent, takes place at the interface between both media. From the point of view of the mill base, a solvent extraction takes place. This diffusion gives rise to a loss of solvent from the mill base and, consequently, to an early precipitation of the binder with the simultaneous formation of conglomerates built up of polymer and pigment particles. This process is schematically illustrated in Figure 4.17 above).

Pigment shock is only reversible when the polymer layer adsorbed on the surface of aggregates and primary particles in the mill base ensures sufficient stabilisation for the conglomeration to behave as a form of flocculation. In this case, the clusters can be redispersed by submitting the let-down system to intensive stirring.

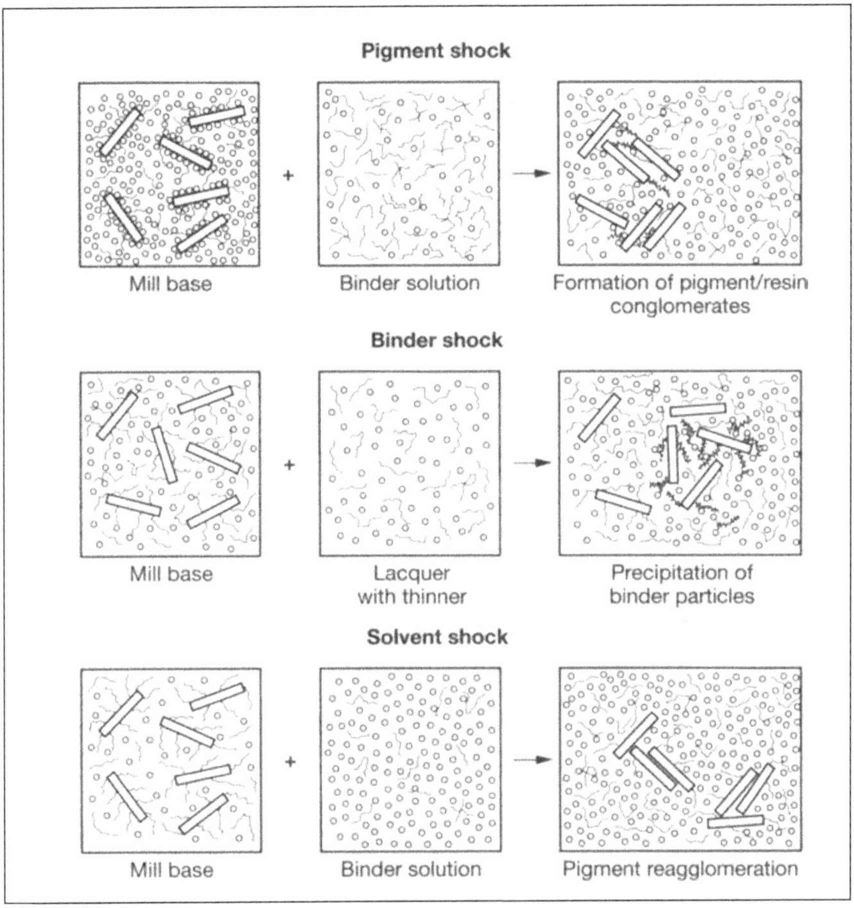

Figure 4.17: Schematic representation of shock effects: pigment shock (above); binder shock (centre): solvent shock (below), see text for further explanations

To avoid this trouble, the difference in binder concentration between mill base and let-down varnish should be minimised as far as possible. On the other hand, care must be taken to formulate the mill base with the solvent having the weaker solving power and the lower surface tension. The first measure hinders diffusion and the second favours the concentration balance.

4.3.3.2 Binder shock

Binder shock is observed when the let-down varnish contains a thinner with only limited solving power and its critical concentration is exceeded during the letting down stage. This leads to binder precipitation in the form of specks which can also entrap pigment particles. The result is irreversible damage to the system. Such a case may also arise if letting down is carried out with a varnish containing a resin only partly compatible with the dispersion resin. The mechanism of binder shock is shown in Figure 4.17, centre.

Binder shock can be prevented by replacing the thinner in the varnish by true solvents and/or by using resins that are mutually compatible.

4.3.3.3 Solvent shock

This phenomenon arises when a pigment dispersion is let down with a solvent-rich binder solution or diluted with pure solvent, and results from the extraction of binder from a pigment. The solvent molecules, rapidly penetrating the mill base, strip the pigment particles of their stabilising polymer layer. Pigment re-agglomeration ensues and nullifies the dispersion effect (schematic in Figure 4.17 below). This change is therefore largely irreversible. Carbon blacks and some finely-divided organic pigments are particularly sensitive to shock.

A remedial measure to counteract binder extraction is primarily a better adjustment of the resin concentration in the varnish to that in the mill base. Besides, an experienced technician would not contemplate letting down with pure solvent.

Practical advice for favourable let-down conditions

A cause common to all three of the shock effects discussed is a high concentration gradient between the mill base and thinner or let-down varnish. The following recommendations should thus be followed when letting down a pigment dispersion:

- A pigment-rich mill base should never be diluted with pure solvent.

The concentration gradient of the resin between mill base and let-down varnish has to be minimised as far as possible.

In order to shorten the contact time of the different phases, the let-down varnish should be added gradually under vigorous stirring conditions.

Always add the varnish to the pigment dispersion and not the other way round.

When using solvent combinations, the mill base should be formulated with the solvents having the lower surface tension and the weaker solving power.

The topic shock effects is discussed in papers by *Haug* [126] and *Sander* [127], as well as in a technical information sheet from *Bayer* [128].

4.3.3.4 Seeding

The phenomenon known as **seeding** results from the formation of metallic soaps by the chemical reaction of zinc white, zinc oxide or other basic pigments with the saturated fatty acids of the binder. Strictly speaking, this effect is **chemical seeding**, in contrast to **physical seeding** as the shock effects discussed above have been termed. For the sake of clarity, it would be advisable to dispense with the expression physical seeding, substitute shock effects for it and adopt the term seeding to denominate only chemical seeding. This would render a confusion of terms improbable.

This phenomenon is characterised by large, seed-like specks (up to approximately 100 µm), which are formed during storage of the pigmented system. These specks are soaps of palmitic or stearic acid, and they often entrap pigment particles on precipitation.

Seeding can be prevented primarily by careful selection of the binder. There are various specific types among the modified alkyd resins, e.g. long-oil types with 60 % oil content or more, that are compatible with zinc oxides. Soya-bean oil alkyds with high contents of saturated fatty acids are, on the contrary, assumed to be particularly capable of forming zinc soaps. As zinc-soap formation is promoted by high temperatures, dispersion should be carried out in mill types allowing temperature increase to be kept within limits, e.g. triple-roll or ball mills. Furthermore, due to the better solubility of metallic soaps in aromatic hydrocarbons, these should be used in the formulation as far as possible instead of aliphatic hydrocarbons.

A detailed discussion of the seeding problem can be found in a study by *Liehr*[129].

5 Dispersion process: methods and mathematical models for assessing the degree of dispersion

5.1 Introduction

The methods discussed in Section 1.3 for particle size research are – with the exception of sieve analysis – less useful for determining the degree of dispersion during production. For this reason, a series of other methods have been developed that either **directly** or **indirectly** register the particle size. The direct methods of measurement include measurement of the fineness of grind, also called coarseness, and observation with the optical microscope. The indirect methods are based on the evaluation of properties of the pigmented system, such as gloss, tinting strength, hiding power or semi-transparency, that are dependent upon the particle size. Common to all methods is the stepwise determination of the dispersion level attained in the mill base with increasing energy input, i.e. by taking samples at specified stages of the dispersion. This can be expressed, for example, as time in the case of bead, sand or ball mills, or number of passages over a triple-roll mill.

As the direct methods provide particle size values down to about 2 µm (fineness-of-grind gauge or Garmsen testing apparatus), in the best case 0.2 µm (optical microscope), their practical information content is restricted solely to the portion of coarse grain still present in the mill base, i.e. the zone limiting the right-hand side of the area under the particle size distribution curve (Figure 1.8b). These methods, however, do not supply any information regarding the degree of dispersion of the fraction below 1 µm – information crucial for important application properties.

5.2 Direct methods

5.2.1 Fineness-of-grind and dispersion curve

Fineness of grind or **coarseness** are terms given to a size value assessed under specific conditions. This size refers to the point at which agglomerates start to

Juan M. Oyarzún: Pigment Processing

ISBN 978-3-86630-665-3

become clearly visible through the surface of a drawn-down sample. The test can be carried out with either the *Hegmann* fineness-of-grind gauge or with the *Garmsen* device.

Hegman system

The fineness-of-grind gauge, also termed grindometer or simply grind gauge, consists of a steel block into which one, sometimes two, wedge-shaped channels have been cut. These channels may, depending on the design, be 100, 50 or 25 µm at the deeper end and they run lengthways to zero depth at the other end. The channel depth can be read as micrometres or as Hegman units from a linear scale engraved alongside the channel or on the side of the block. A sample of the dispersed mill base is placed in the deeper end of the channel and drawn down with a scraper. Any oversize pigment particles present protrude through the surface of the wet film at that point where their diameter exceeds the channel depth, and they can be seen as specks or scratches. When reading off the value determined, isolated specks are ignored and the value on the scale corresponding to the point at which visible particles sharply increase in number is noted.

The conversion of Hegman units into micrometres can be performed by means of the relationship:

$$\text{Hegman units} = 8 - (1/12.7) \cdot x \text{ [value in micrometres]}$$

When using the grind gauge, it is important to know that the channel is by no means completely filled with paint, so that the depth of the paint wedge does not correspond to that of the channel. A series of measurements has shown that the filling degree of the channel may vary between 70 and 85 %. The value recorded does not reproduce the actual facts. This shortcoming makes no difference in those cases where pigment dispersions in one and the same medium are being compared, as the deviation is then constant. A comparison of values obtained with samples from different binder/solvent systems would be, however, at least questionable.

It should also be pointed out that grindometer values below 5 µm are generally dubious due to solvent evaporation. This is particularly the case when the more volatile solvents outweigh in the sample composition. In addition, it may occasionally arise that the sample wedge in the channel shows a relatively long transition zone between isolated specks and the beginning of the main agglomerate population. Determination of a reliable fineness-of-grind value is impossible in such cases.

Reproducible values can generally be obtained with the grind gauge if certain precautions (as stated in DIN EN ISO 1524) are taken when carrying out the measurement. The assessment of dispersion characteristics from the fineness of grind is specified in the standards DIN EN ISO 1524, ISO 8781-2 and ASTM D 1210.

Garmsen method

Measurement of coarseness with the Garmsen device is based on the same principle. The apparatus comprises two accurately ground, glass plates with a lateral scale from 1 to 10 cm. A shim of suitable thickness (200 µm shims are standard) is laid across the end of the plate marked with 10, the sample is poured onto the plate close to the shim and, finally, the second plate is placed on top. Both glass plates are then clamped together with two bulldog clips. The clamps are removed after approximately 1 minute and the plates separated by sliding the top plate to the side, keeping it parallel to the contact surface. Due to the shim, a wedge-shaped layer of the sample –half on each plate – is obtained, and the layer tapers by a twentieth of the chosen shim thickness from one mark to the next. Reading is performed in the same manner as for the grindometer, i.e. isolated protruding specks are not taken into account. The *Garmsen* procedure demands slightly more effort but is, in turn, slightly more accurate than the grindometer method. A detailed description of the device and its handling have been published in a paper by the inventor [130].

Further methods of assessment

The shortcomings of coarseness measurement using the grindometer or the Garmsen method led to the development of an optoelectronic grindometer. The device consists of a movable grindometer, and optical and electronic systems. A laser beam irradiates the wedge-shaped sample in the channel and is scattered by the pigment specks present. The scattered radiation is amplified, recorded by an oscilloscope and transmitted to a computer for evaluation. The result is shown either as impulses of scattered light in the oscilloscope or illustrated as a size distribution in a histogram. This method avoids the influence of subjective factors, as may occur in the case of grindometer measurements. In the inventor's opinion, the values obtained with the optoelectronic grindometer are very reproducible as long as air bubbles are avoided and measurements are carried out under dust-free conditions [131].

Evaluation system

Plotting the recorded set of coarseness values against milling time or passages, i.e. indirectly against energy input, in a *Cartesian* diagram generally results in the curve of a decreasing monotonic function. The graph at the left-hand side of Figure 5.1 shows the curve best fitting the values obtained by milling a red iron oxide pigment (C.I. Pigment Red 101, N° 77941) in an aqueous alkyd resin system. This form of illustrating the dependence of the fineness of grind upon milling time is termed a **dispersion curve**. A good linear relationship between fineness of grind and energy input is usually found by plotting the values on double log paper, as shown by the graph at the right-hand side of Figure 5.1.

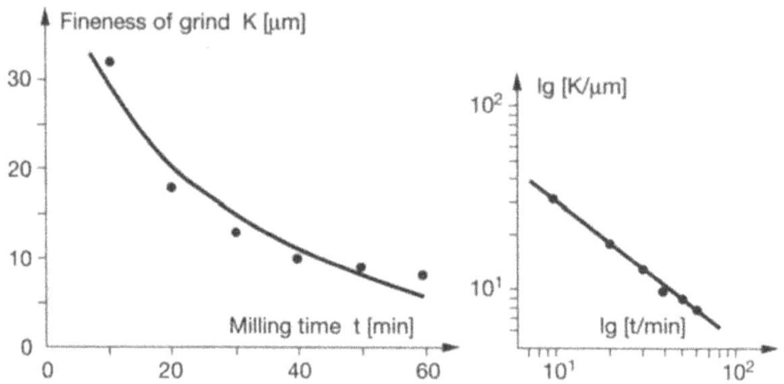

Figure 5.1: Decrease in coarseness with increasing dispersing time for an iron oxide pigment in linear (left) and double logarithmic plot (right)

The resulting straight line is expressed by the equation:

$$\log K = \log \mathrm{A} + \mathrm{B} \log t$$

where K is the fineness of grind, t the milling time and A and B system-dependent constants. Hence, for the time-dependence of the fineness of grind it follows that:

Equation 5.52:

$$K = \mathrm{A} \times t^{\mathrm{B}}$$

The fineness of grind at the beginning of the milling process is given by constant A, which is obtained by linear regression or by graphical extrapolation to $\log t = 0$ (i.e. $t = 1$ time unit). Constant B is a measure for the mean velocity of the decrease in coarseness. Model Equation 5.52 provides a good description of the mathematical relationship between fineness of grind and energy expenditure. An important consequence for the practice is that two parameters are necessary to express the process numerically [132].

Following the dispersion progress by measurement of the coarseness is not possible in the case of finely-divided or easily-dispersible pigments. The coarse fraction will be found to be below the resolution limits of the grind gauge after only a short milling time, although the intended degree of dispersion is still far from the point to be reached.

An interesting paper by *Schwegmann* considers the pros and cons of fineness-of-grind measurements [133].

5.2.2 Examination with the optical microscope

Due to its limited resolving power, the optical microscope is unsuitable for detecting particles smaller than the wavelength of light, i.e. below 380 nm. However, when merely an impression of the coarse fraction is required, examination with the optical microscope may be a good complement to fineness-of-grind measurements.

Kellner and *Weyres* combined both methods to attain a microphotographic evaluation of grindometer draw-downs. According to their proposal, a specific area of the sample wedge on the grind gauge is photographed at twenty-fold magnifica-

Coarseness pattern

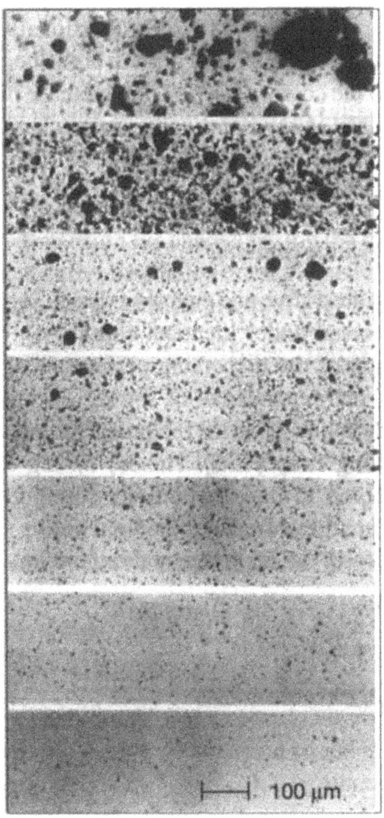

Scale	Particle size
1	over 100 µm
2	to approx. 100 µm
3	to approx. 50 µm
4	to approx. 25 µm
5	to approx. 10 µm
6	to approx. 5 µm
7	below 5 µm

Figure 5.2: Goetze scale for visual assessment of the fineness of grind

Source: BASF SE

tion under defined conditions of illumination, and the specks appearing on the photomicrograph are counted [134]. The method requires more effort than coarseness measurements or microscopic examination separately, but is, in turn, more precise. It is less to be recommended for controlling production batches than for basic research in, for example, rating the efficiency of different types of dispersion equipment. One of its main advantages is that it permits the storage of photomicrographs for later comparison with others resulting from subsequent tests.

A further microscopic method that can be recommended for rapidly controlling the dispersion process in paint or ink manufacture is evaluation of the coarse fraction by means of the *Goetze* scale. This scale, shown in Figure 5.2, consists of seven photomicrographs in the ratio 1:80 with the coarseness decreasing from picture 1 (>100 μm) to picture 7 (< 5 μm).

The mill base sample is drawn down on a slide and observed under transmitted light at 80-fold magnification. A quantitative rating is not given in this method. The level of dispersion is stated by comparing the overall appearance of the sample under the microscope with the scale.

The microscopic evaluation of coarseness for pigment and filler dispersions according to the *Goetze* method has been taken as the basis for a DIN standard, although the scale used only comprises six steps [135].

5.3 Indirect methods

5.3.1 Tinting strength development

Among the indirect procedures for assessing the degree of dispersion, determination of the tinting strength development is very widespread due to its clarity and practical importance. As the increase in colour strength is not only dependent on the dispersion energy, but also upon the binder/solvent system and the dispersing equipment, test methods for tinting strength development in a series of media and mill types are specified in various standards (e.g. DIN EN ISO 8781-1).

Essentially, mill base samples taken in accordance with agreed dispersion stages, e.g. time, number of revolutions or passages, are reduced with a white stainer and their depth of shade is colorimetrically measured after application and drying of the reductions. Care must be taken at this stage to ensure that no separation processes falsify the result. This can be simply verified with the aid of the rub-out test. The tinting strength development is thus followed indirectly on the basis of depth of shade. The K/S-values, calculated from the reflectance at the absorption maximum using the *Kubelka-Munk* relationship (Equation 2.20), are used as a measure for the tinting strength. As an example, Figure 5.3 (left) shows the

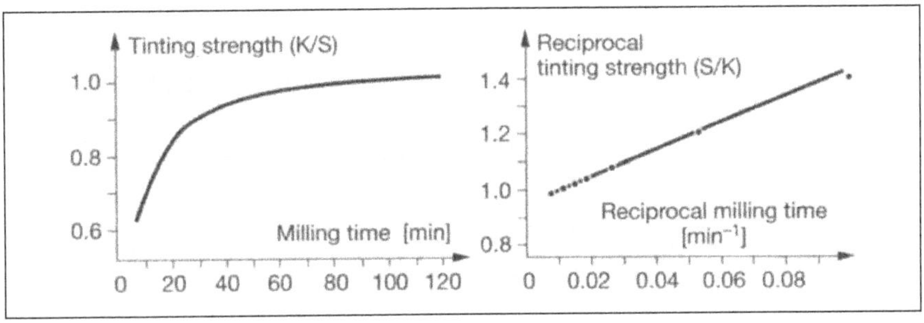

Figure 5.3: Tinting strength development of a copper phthalocyanine blue pigment in an aqueous paint system in linear (left) and double reciprocal plotting according to von Pigenot (right)

tinting strength development of a copper phthalocyanine pigment dispersed in an aqueous paint system.

As a rule, the tinting strength of a pigment does not rise proportionally to the energy input during the dispersion process. After a rapid growth at the beginning, due to the relatively easy break-down of the coarser fraction, the increase in tinting strength becomes slower and slower until it asymptotically approaches a limit, denoted **final tinting strength** F_∞, that can be theoretically attained after infinitely high expenditure of energy for the dispersion.

Tinting strength development as a hyperbolic function

The strength development curve is very illustrative and enables a direct comparison between pigments that have been dispersed under identical conditions, but it is inadequate for a quantitative description of the dispersing performance. *D. von Pigenot* was able to demonstrate that the dependence of tinting strength upon milling time can be represented as a straight line when the reciprocal tinting strength is plotted against the reciprocal milling time [136]. Hence, tinting strength development can be formally considered to be a hyperbolic function of the dispersion expenditure. Its asymptotes are the ordinate of the final tinting strength and the *y*-axis or a parallel to it. Therefore, the mathematical expression for this relationship is:

Equation 5.53:

$$\frac{1}{F} = \mathrm{a} + \mathrm{b}\frac{1}{E}$$

where *F* is the tinting strength, indirectly measured as *K/S* (depth of shade), *E* is the energy input (e.g. milling time, number of passages on a triple-roll mill etc.) and a and b are system-dependent constants.

In the linear representation, illustrated on the right in Figure 5.3, constant a is given by the intersection of the extrapolated straight line with the ordinate axis. Its reciprocal thus corresponds to the final tinting strength F_∞. Constant b, the slope of the straight line, is a measure for the **mean rate of dispersion**, i.e. for the average value of the ratio **increase in tinting strength : energy input**. The steeper the straight line, the higher is the mean rate of dispersion. However, constant b is inconvenient for a quantitative characterisation of the dispersion process because pigments having markedly different initial and final tinting strengths, and being also completely unlike as to their dispersibility, may nevertheless have equal mean rates of dispersion. *Von Pigenot* proposed the use of the segment formed on the abscissa by extrapolating the straight line to its intersection with the x-axis as a suitable dimension for characterising the dispersion process. Its reciprocal has the unit of a negative dispersion energy. The following relationships for this dimension result from the schematic graph in Figure 5.4:

$$b = \tan \alpha = a \cdot E_o$$

$$E_0 = \frac{b}{a} = \frac{\frac{1}{F_n} - a}{\frac{1}{E_n}} \left(\frac{1}{a}\right)$$

This equation can be reduced to the form:

Equation 5.54:

$$E_0 = E_n \left(\frac{F_\infty}{F_n} - 1\right)$$

In this relationship, E_0 is the dimension that gives the energy input, thus characterising the dispersion process quantitatively; E_n corresponds to the energy input required for an arbitrary degree of dispersion during the process, F_n is the tinting strength obtained at this stage, and F_∞ the final tinting strength. Equation 5.54 will be reconsidered in Chapter 6 in connection with dispersibility.

Equation 5.53 expresses the hyperbolic dependence of the tinting strength upon the energy input, but neglects the initial pigment tinting strength F_0, i.e. the value actually present before dispersion. In practice it is credited to the dispersion expenditure, which results in pigments with high initial tinting strength simulating ease of dispersibility even when considerable dispersion expenditure is required to develop the remaining tinting strength. This is apparent from the

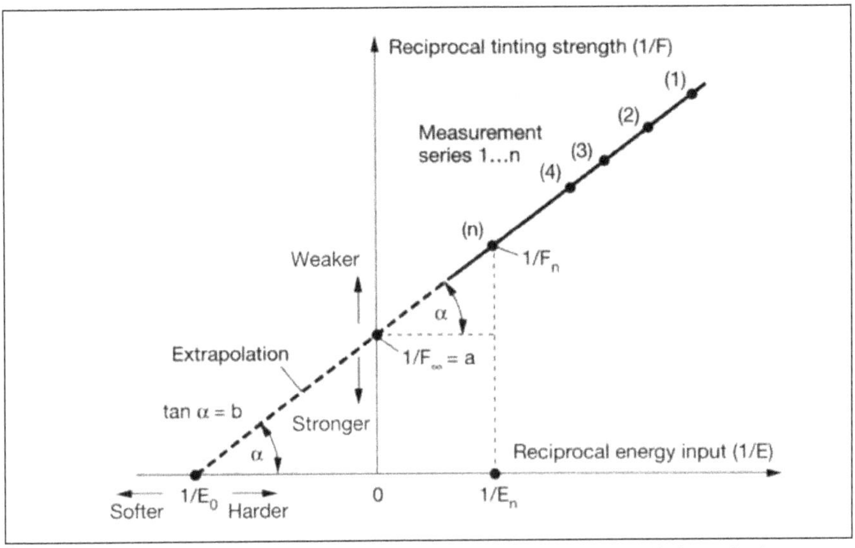

Figure 5.4: Model system for calculation of the dispersing effort with the aid of a hyperbolic function according to von Pigenot

linear plot by the fact that the values of the reciprocal tinting strength in the upper zone, i.e. at the initial stage of the process, lie below the straight line and their distance to the line becomes shorter and shorter with progressing dispersion. *Von Pigenot* noted these facts and suggested a description of tinting strength development by means of an exponential function as an alternative. However, he was of the opinion that the hyperbolic representation gives a correct reproduction of the stage of development that is important to the pigment processor, namely the last third, and this has been confirmed in practice.

Tinting strength development as an exponential function

An exponential function as a model for the tinting strength development was worked out by *O. J. Schmitz* et al. [137]. They proceeded from the hypothesis that the pigment powder consists solely of primary particles and agglomerates, and that the latter are immediately split into primary particles during dispersion. In addition, the interaction between agglomerates was neglected, and it was assumed that tinting strength is caused only by the primary particles. Under these prerequisites, which simplify the actual facts, the decrease in agglomerates per unit time must be proportional to the number of agglomerates present at time *t*. This can be expressed by the following differential equation:

$$-\frac{dN}{dt} = k\,N$$

This equation may also be written:

$$\frac{dN}{N} = -k\,dt$$

Introducing the limits of integration, i.e. N_0 for the time $t = 0$ and N_t for the agglomerates dispersed at a later point of time t, we obtain:

$$\int_{N_0}^{N_t} \frac{dN}{N} = -\int_0^t k\,dt$$

The integration yields to Equation 5.55:

$$N_t = N_0 \times e^{-kt}$$

The number N_0 of agglomerates present at time $t = 0$ must equal the difference between the total number of agglomerates N and the number of agglomerates already broken down at time $t = 0$, which we term N_Z. Thus:

$$N_0 = N - N_Z$$

On the other hand, the sum of the agglomerates dispersed at time t plus the broken down agglomerates (i.e. $N_{(D+Z)} = [(N_0 - N_t) + N_Z]$) must equal the difference between the total number of agglomerates N and the number of agglomerates still present at time t. Consequently:

$$N_{(D+Z)} = N - N_t$$

By introducing the expression for N_t from Equation 5.55, and replacing N_0 by the difference $N - N_Z$, we obtain Equation 5.56.

Equation 5.56:

$$N_{(D+Z)} = N - (N - N_Z) \times e^{-kt}$$

Since the tinting strength is directly proportional to the number of dispersed agglomerates, Equation 5.56 may also be written:

Equation 5.57:

$$F_t = F_\infty - (F_\infty - F_0) \times e^{-kt}$$

where

F_0 = the initial tinting strength at time $t = 0$,
F_∞ = the final tinting strength at time $t = \infty$,
F_t = the tinting strength at time t,
k = a constant.

Slightly modifying Equation 5.57 and expressing it in logarithmic form gives Equation 5.58:

Equation 5.58:

$$-\ln\left(1-\frac{F_t}{F_\infty}\right) = -\ln\left(1-\frac{F_0}{F_\infty}\right) + k \cdot t$$

The final tinting strength F_∞, which theoretically corresponds to the dispersion time $t = \infty$, remains unattainable in practice. However, after a sufficiently long, finite dispersion time it is possible to obtain a tinting strength that only differs insignificantly from the final strength. If this tinting strength is given the value 100, the F_t-values determined in the course of the dispersion process may be expressed as percentages of the tinting strength considered as final.

Plotting $-\ln[1 - (F_t/F_\infty)]$ against the dispersion expenditure or t in Cartesian co-ordinates results in a straight line. Its extrapolation to the ordinate axis gives the ordinate segment $-\ln[1 - (F_0/F_\infty)]$. The slope corresponds to constant k, which represents a measure for the mean rate of dispersion. It is thus possible to quantitatively describe the tinting strength development by means of the three parameters F_0, F_∞ and k. Figure 5.5 illustrates the linear representation of a copper phthalocyanine green pigment (C.I. Pigment Green 7, N°. 74 260) dispersed in linseed oil varnish on a laboratory triple-roll mill as an example for such a graph.

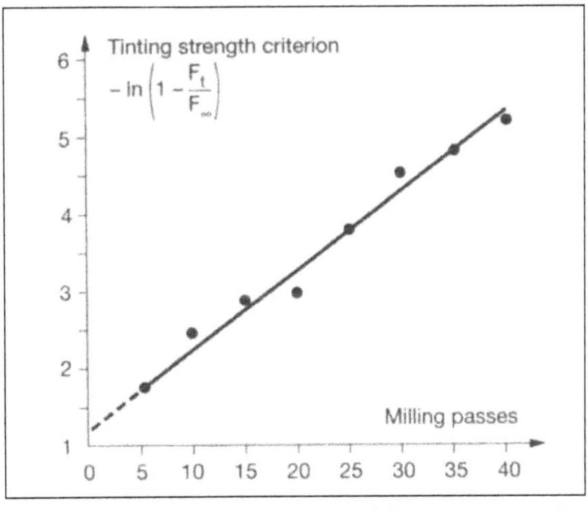

Figure 5.5: Linear transformation of the tinting strength development curve according to the exponential function of Schmitz et al.

Comparative appraisal of both methods

In a comparison between the two procedures for quantitatively characterising tinting strength development, the **hyperbolic** function of *von Pigenot* and the exponential function according to *Schmitz* et al., *Zorll* pointed out that in the first case the hyperbolic dependence of the tinting strength upon the dispersion time can be explained by considering the process to be a second-order reaction. The differential equation stated as a starting point in the second case corresponds to the type of equation used in chemical kinetics to describe first-order reactions [138].

Interpretation of the dispersion process as a second-order reaction is justified by the fact that the agglomerates' own milling effect makes an additional contribution to the tinting strength development. This state of affairs would contradict an approach to the tinting strength development in the sense of an exponential function, as this does not consider the milling effect of the agglomerates. Nevertheless, application of the exponential function in practice has stood the test of time. This may probably be attributed to processes counteracting the agglomerates milling effect, e.g. re-agglomeration, and therefore lessening its influence. In any case, the mutual interaction between the agglomerates strongly depends on pigment and on milling conditions.

In a series of tests confined solely to **inorganic** pigments, *Klaeren* and *Völz* found that tinting strength development could, in most cases, be better described by the hyperbolic function. However, some cases were found in which both functions proved to be equivalent [139].

Tinting strength measurement during dispersion

A method has been recently described for determining tinting strength development in the course of dispersion. The process takes place in a normal bead mill, and the mill base is transported by a pump through a secondary circuit to the sampling point where a second pump provides for accurate sampling. Dosage of a white stainer to the samples is carried out by a third pump. After thorough mixing with a stirrer, measurement is performed through a specially designed window using a commercial colorimeter. The data obtained are evaluated by a computer connected to the system.

The procedure and arrangement were described by *Lust* [140]. This method has the great advantage of enabling the dispersion operation to be interrupted as soon as the tinting strength only shows insignificant further increases, or before secondary negative processes, such as overgrinding or re-agglomeration, gain the upper hand and cause the tinting strength to decrease. It remains to be seen how this method stands the test of production practice.

5.3.2 Assessment of the state of dispersion from gloss development

When discussing gloss and gloss haze in Section 2.1, it was also mentioned that the development of gloss may serve as a criterion for assessing the degree of dispersion. The surface of a dry coating, even when it only consists of a thin, clear binder layer, has an uneven structure caused by pigment particles lying close beneath it. The larger the particles, the more pronounced must be the surface structure, independent of any possible protrusion of single particles. As the specular reflection of a coating depends directly on its surface structure, the close relationship between gloss and degree of dispersion thus becomes clear. Two detailed studies by *Braun* and *Fields* deal with the mechanism by which pigment particles affect the gloss of paint films [141, 142].

The determination of gloss development may be carried out in accordance with a method specified in the standards DIN EN ISO 2813 and DIN EN ISO 8781. The goniophotometer or reflectometer values obtained in each case are plotted against the dispersion expenditure (e.g. dispersion time) in a Cartesian diagram on either a linear or on a double reciprocal scale in the same manner as for tinting strength development. However, it is in this case not necessary to use the slope of the straight line as a measure; the dispersing work required to attain a given gloss value can rather be obtained by extrapolation. Figure 5.6 shows the graphical representation, in a linear and in a double reciprocal plot, of the gloss development of an inorganic mixed-phase

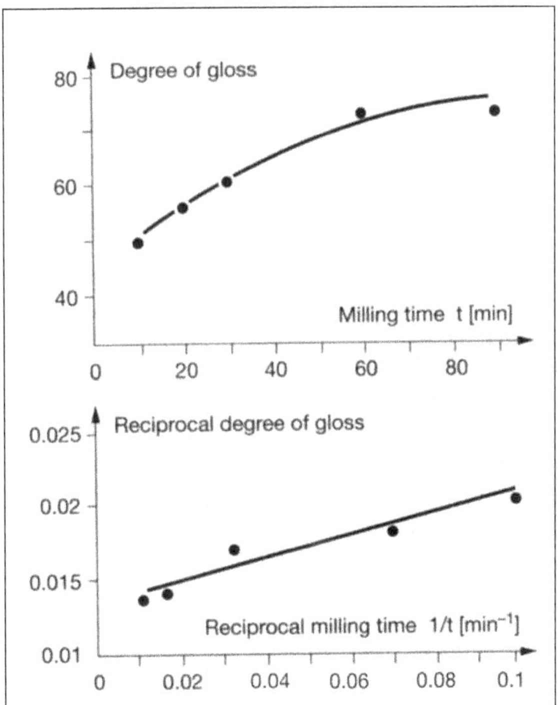

Figure 5.6: Gloss development of a mixed phase pigment in an alkyd/melamine storing system in linear (above) and double reciprocal plotting (below)

pigment (C.I. Pigment Brown 24, N° 77 310) dispersed in an alkyd/melamine stoving finish.

When comparing the dispersion behaviour of different pigments with the help of gloss measurement of coating layers produced with them, it should not be forgotten that the type of medium and pigment volume concentration also have an influence on gloss. Thus, for example, gloss measurements of air-dried, long-oil alkyd based paints gave rise to different values for the same pigments at equal PVC in an alkyd/melamine stoving finish [143]. According to *Jefferies,* a possible explanation for that result is the shrinkage of the stoving finish film due to condensation of the melamine resin [144].

5.3.3 Increase in lightness flop

The characteristic feature by which metallic-effect coatings (usually pigmented with aluminium powder) differ from other effect or uniform finishes is the phenomenon known as **flop**. This is the term used for the change in hue and/or lightness of a coating layer depending on the angle of view and the direction of incident light. The change in lightness, termed **lightness flop**, can be ascribed to the particularly high reflectance occurring in the neighbourhood of the gloss angle at the expense of reflectance in other zones. Thus, the more the angle of view deviates from the gloss angle, the more pronounced is the decrease in reflectance, i.e. the coating layer will appear darker to the observer.

If the metallic-effect coating also contains a coloured pigment, the light reflected by the aluminium pigment is selectively absorbed and, on the other hand, scattered by the coloured pigment, the latter effect depending on the scattering power of the particles. The compositions of the scattered and of the transmitted spectra are also directionally dependent on one another, i.e. the observer also perceives a change in hue with varying angle of view. This is termed **hue flop**. The combined effects of hue and lightness flop give rise to the phenomenon known as **goniochromatism** in metallic effect coatings or in transparent finishes on strongly reflecting substrates.

Whereas the hue flop is always dependent on the effect of scattering particles, lightness flop may also occur in coating finishes produced with transparent pigments or dyestuffs. When scattering pigment particles are present in the coating system, the luminance rises in the scattering zone and lessens in the reflecting zone, causing the lightness flop to decrease. Thus, the lightness flop brought about by transparent pigments must increase with progressing dispersion, and the development of lightness flop may be used as a criterion for the degree of dispersion of **transparent** pigments [145].

The measurement principle for determination of lightness flop may easily be derived from the foregoing explanations. The reflectance of the coating finish is measured under two different angles, chosen preferably in the reflecting and in the scattering zones. The corresponding lightness values are calculated from the data obtained, and the difference is the lightness flop. The CIELab system, described in DIN EN ISO 11664-4 for the determination of colour difference is very adequate for this purpose. A method proposed by *Finus* uses an automated goniophotometer that allows continuous measurement of the lightness flop over its whole range of angular dependence [146].

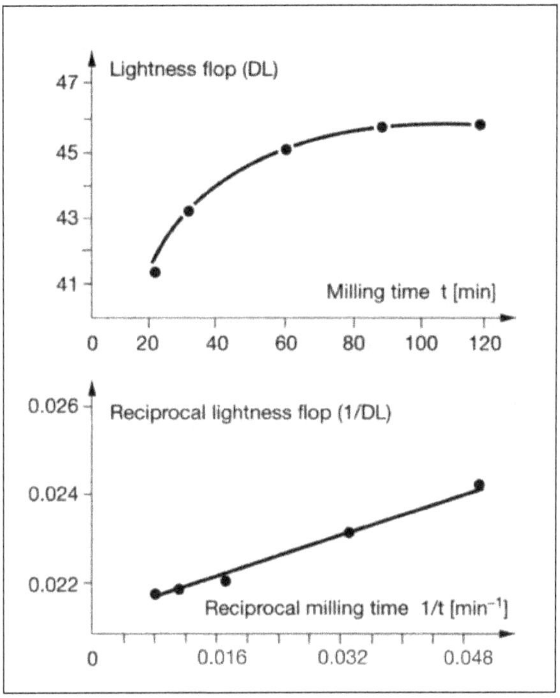

Figure 5.7: Development of lightness flop for a transparent iron oxide pigment as a function of dispersing time: linear plot (above) and linear transformation by double reciprocal plot (below)

Figure 5.7 above illustrates the lightness flop values DL (**difference in lightness**) in CIELab units (DIN EN ISO 11664-4) as a function of the dispersion time t (in minutes) for a transparent iron oxide (C.I. Pigment Yellow 42).

The data obtained from a series of tests were analysed by computer in order to find the best mathematical model for characterising the course of the curve. From the thirteen possible equations that were convertible into linear form, the best fit was:

Equation 5.59:

$$DL = \frac{t}{a + b\,t}$$

It could be stated that this relationship delivered the best correlation coefficients, in some cases virtually 1. A slight transformation of Equation 5.59 yields the equation of a straight line.

$$\frac{1}{DL} = a\frac{1}{t} + b$$

Hence it is obvious that by plotting the reciprocals of *DL* against the reciprocals of *t*, the curve representing the lightness flop development can be converted into a straight line, as in the case of the tinting strength development (Figure 5.7, bottom). It is thus possible to easily interpolate the value of the milling time required for a given lightness flop. Constants a and b are system-dependent [147].

5.4 Concluding remarks

Summing up at the end of this discussion of methods for assessing the degree of dispersion in production, we may state the following:

It is important for the pigment processor to know that, when comparing different pigments with regard to their dispersion behaviour, the result also depends on the evaluation criterion chosen. Following the dispersion progress by grindometer measurements is quick and inexpensive, but the information remains confined to the coarse fraction in the mill base. Only indirect methods, such as the development of gloss, tinting strength or transparency, are capable of providing data on the state of particle size reduction in the fine distribution zone.

When dispersing inorganic pigments, particularly full-shade pigments, the optimum optical properties are as a rule attained relatively rapidly, whereas the decrease in the coarse fraction proceeds more slowly. Direct methods are thus generally more appropriate for these kind of pigments. Organic pigments and carbon blacks, particularly the finely-divided grades, behave in the reverse fashion. Assuming adequate dispersion conditions, their coarse fraction generally lies beyond the resolution limits of the grind gauge after only a relatively short milling time. However, they require a longer period of energy input to develop their optical properties fully. It would thus be wrong in their case to stop the dispersion process when coarse fractions can no longer be determined on the grind gauge or the Hegman device. For this reason, **indirect methods** are ordinarily to be preferred in the case of organic pigments.

Naturally, in addition to the methods discussed here, several other methods may also be taken into consideration where use is made of alternative properties of the pigmented system as criteria for the evaluation of the level of dispersion, e.g. transparency, hiding power or viscosity. Which criterion is to be given preference depends, in the final analysis, on the intended use of the final product.

6 Dispersion process: mill base optimisation

6.1 General considerations

From a purely physical standpoint, liquid pigmented systems, such as paints, enamels and printing inks, are suspensions of pigments finely dispersed in an external continuous phase that may consist of oils, varnishes, plasticizers, resins, resin solutions etc. The optical, mechanical and other properties of the final product are determined not only by its components and their concentrations in the system, but also to a high degree by the state of pigment dispersion attained in the dispersing process. Hence, a successful dispersion is just as important for the pigment processor as the correct selection of the system components and the exact formulation of their amounts in the finished product. On the other hand, the energy input required for the dispersion, and thus its share in the production costs, is also dependent upon the dispersing conditions. Thus, by carefully adjusting the factors that govern the dispersion process, it is possible to obtain the best result with regard to both quality and costs of the final product.

These factors may be classified into the following groups:

1. Chemical constitution, physical properties of pigments, binders and solvents, and their interaction
2. Type and amount of dispersing agents
3. Mill base preparation
4. Type and mode of operation of the dispersing equipment
5. Pigment, binder and solvent concentration
6. Mill base viscosity
7. Temperature
8. Milling time
9. Milling energy input

The selection of the materials listed under group 1, i.e. pigments, vehicles and solvents, is generally dictated by the desired properties and intended purpose of

Juan M. Oyarzún: Pigment Processing

ISBN 978-3-86630-665-3

the finished product, so that in this case the pigment processor has only little or no freedom to influence the dispersing operation. He may possibly improve the dispersing conditions by choosing the easily-dispersible or micronized grades of the pigment, if available, or by carrying out the dispersion in the vehicle with the better wetting properties if the system contains two or more types of vehicle. An appropriate solvent combination may also help, but even this possibility is often limited by other aspects.

Dispersing agents

As a rule, the addition of surfactants – termed **dispersing agents** in this context – can enhance the dispersing effect. However, there are also surfactants that may negatively influence the dispersion process [148]. As mentioned in Section 4.3.2, the variety of dispersing agents and other raw materials suitable for consideration as components for the mill base do not permit the establishment of specific guidelines for this area. The pigment processor must determine type and concentration of the dispersing agent suitable for his system by carrying out his own tests based on the information provided by the supplier.

Mill base preparation

Preparation of the mill base plays an important role. Numerous pigments show a distinctly better dispersion performance, when **wetting-out** has taken place before grinding. Wetting-out is the term used for the storage of the mill base for a period (e.g. overnight) before the actual dispersion process begins. During this stage wetting continues, slowly, so that part of the agglomerates can be broken down without the need of mechanical work. Energy expenditure, coupled with an increased temperature when adding the pigment during premixing as well as during mill base homogenisation, also have a favourable influence on the subsequent dispersion.

Dispersing equipment

The type of dispersing machine to be used is determined primarily by the mill base consistency. Roll mills or high-speed disc impellers, such as for example the *Cowles* dissolver type, are the proper equipment for high-viscosity pigment pastes. Relatively low-viscosity systems are processed batch-wise in ball mills and attritors, and continuously in sand or bead mills. A further criterion for choice of the appropriate type of dispersing equipment is the quality of the final product. For example, dispersion with a dissolver is often sufficient for plasters and other house paints, particularly when micronized or easy-dispersible pigments are used. On the other hand, high-quality industrial or automotive finishes, which must fulfil high demands as to gloss and fineness of grind, will require the use of machines with a higher dispersing effect, e.g. triple-roll mills, ball, bead or sand mills. In particular cases, such as surface-treated pigments or pigments sensitive to milling, the experienced operator will always select machines, such as the dissolver or triple-roll mill, which have a more gentle milling effect.

The mode of operation of modern dispersing machines is the subject of thorough studies by *Herbst* [149–152]. In a paper by *Roelofsen* the limits of stirrer-type ball mills are discussed [153]. A compilation by *Ruf* deals with operation and use of the main dispersing machines [154].

Effect of temperature

The influence of temperature has been dealt with in the preceding chapters in connection with viscosity and wetting (see Sections 3.3 and 4.2.2). Most pigments can be better dispersed at raised temperature (generally 40 to 70 °C), provided that this is tolerated by the binder system. Although wetting is an exothermic process, which is therefore favoured by the removal of heat, an increased temperature counteracts the increase in viscosity and causes a reduction in surface tension, and thus the positive effects predominate. An elevated dispersion temperature only has a negative influence in the case of systems in which the breakdown of pigment agglomerates is performed by shear, as the resulting decline in viscosity is detrimental to this process.

However, there are also organic pigments whose dispersibility is not improved by increased temperature, but, on the contrary, even worsened. In a series of tests, *Herbst* was able to demonstrate that a temperature rise from 20 to 60 °C during the dispersion of dioxazine violet in a toluene-based gravure ink resulted in a distinct decline in tinting strength as well as a change in hue towards blue. A similar effect was noted with a naphthalene tetracarboxylic acid derivative (C.I. Pigment Orange 43) tested in the same medium. The loss in tinting strength was ascribed to pigment recrystallization, i.e. the finely dispersed particles dissolve and grow together again to form larger crystals. Thus, when processing organic pigments in systems containing critical solvents, the temperature should always be kept low if the dispersion is carried out in stirrer-type ball mills or similarly operating machines [152].

The critical pigment volume concentration (CPVC) is also influenced by the dispersion temperature. This concept refers to that particular PVC at which some properties of the coating layer undergo a fundamental change. The amount of binder at the CPVC is just enough to surround and bind the pigment or filler particles. Taking several organic pigments for letterpress and offset inks as an example, *Merkle* and *Herbst* were able to show that a temperature rise improved the level of dispersion and shifted the CPVC towards lower values [155].

Mutual interaction between viscosity and temperature

The remaining factors governing the dispersion process are not independent as they mutually influence one another. At the start of the milling operation the mill base viscosity is primarily determined by the concentrations of pigment, binder and solvents, and by their interaction. However, even if the operator succeeds in keeping the temperature constant, the viscosity does not remain constant during processing. In general, viscosity increases as the dispersion progresses due to the continually increasing immobilization of binder on the surface of the pigment particles, a fact that must be taken into consideration when the mill base composition is worked out. One consequence of this immobilized binder layer is the formation of pigment-binder-pigment bridges, which in their turn give rise to spatial structures resulting in an increase in viscosity and anomalous flow properties. The higher viscosity causes a rise in temperature which, on the other hand, counteracts the increase in viscosity.

Occurrence of anomalous flow properties

With regard to anomalous flow properties, *Daniel* has stated that the mill base should be slightly dilatant. However, a high degree of dilatancy, as well as thixotropy and pseudoplasticity, are unfavourable for the dispersion operation. The degree of dilatancy in organic media can usually be adjusted by varying the binder concentration, and in aqueous systems by the use of additives, e.g. polyphosphates [156].

The influence of various parameters on the outcome of dispersion was the subject of a study by *Hafner*. Using two organic pigments considered difficult to disperse, the dependence of tinting strength development and rheology upon pigment and binder concentration, temperature, type of binder and solvents mixture were examined. One important conclusion was that the terms *easy or* **difficult to disperse** must be considered to be relative, as pigment dispersibility strongly depends on the binder system [157].

Thus, to formulate an optimum mill base, the pigment processor is restricted in practice to variation of the parameters quoted under group 5, i.e. pigment, binder and solvents concentrations. The optimisation of these three factors will be the subject of the next sections.

6.2 Description of the dispersion process by a probability function

6.2.1 Theoretical principles

Some years ago *Winkler*, *Klinke* and *Dulog* succeeded in deriving a probability relationship for quantification of the dispersion process. They started from the hypothesis that two prerequisites, contingent on chance, must be fulfilled simultaneously in order that an agglomerate might be broken down into its structural elements [158, 159].

1. It must come into a volume element in which it can be stressed, i.e. in a potentially active dispersing site.
2. In this volume element, forces should occur that are high enough to break down the agglomerate.

These partial probabilities, that of the geometrical (P_g) and that of the energetic requirement (P_E), are independent of one another and determined solely by the dispersing machine and the mill base composition. Both of them can be expressed by mathematical relationships. Their product gives the total probability: $P_T = P_g \times P_E$.

The dispersion does not take place in the total mill volume, but in temporarily active and constantly changing volume elements. Their sum total represents a fraction of the total volume that is termed effective volumen V_e.

Geometrical partial probability P_g

The geometrical or stress probability is a function of the milling time. Assuming that a number N_n of agglomerates are evenly distributed in the total volume V_T, the number of non-dispersed agglomerates present in the effective volume must amount to $N_n \times V_e/V_T$. Their decrease per unit time is assumed to be proportional to the number of agglomerates present in the **effective volume**.

Equation 6.60:

$$-\frac{dN_n}{dt} = k \cdot N_n \cdot \frac{V_e}{V_T}$$

The proportionality constant k expresses the speed with which the agglomerates are transported into the effective volume, a process that is controlled by the mill base viscosity. Integration of Equation 6.60 within the limits 0 and t leads to:

Equation 6.61:

$$\frac{(N_n)_t}{(N_n)_{t=0}} = e^{-k \frac{V_e}{V_T} \cdot t}$$

Equation 6.61 gives the ratio of the number of agglomerates still present after the milling time t to the number of agglomerates initially present at the time t = 0, i. e.

when the milling operation started. The number of dispersed agglomerates (N_d), must equal the difference between these two amounts.

Equation 6.62:

$$(N_d)_t = (N_n)_{t=0} - (N_n)_t$$

From Equations 6.61 and 6.62 follows:

Equation 6.63:

$$\frac{(N_n)_t}{(N_n)_{t=0}} = 1 - e^{-k\frac{V_e}{V_T}\cdot t} = P_g$$

Equation 6.63 is the mathematical expression for the geometrical partial probability P_g.

Energetic partial probability P_E

The partial probability P_E is determined by the ratio between the effective energy E in the volume element where the agglomerates are presently located, that is the energy density (energy per unit volume, with E/V_T as symbol) and the binding forces holding the structural units of the agglomerates together. As forces with differing strengths occur in the agglomerates, it is convenient to express this force distribution in the form of a mean tensile strength (bond energy) σ, with the same dimension of the energy density. The factor determining P_E is, therefore, the ratio of the energy density to the mean tensile strength, i.e. $E/V_T \times \sigma$.

The ratio of the decrease in the number of the agglomerates present at the beginning of the dispersion process to that energy quantity must be proportional to the number of agglomerates present at any stage of the dispersion, i.e.

Equation 6.64:

$$-\frac{dN_n}{d\left(\frac{E}{V_T\cdot\sigma}\right)} = a \times N_n$$

The proportionality constant a should be considered as an energetic efficiency factor characterising the degree of the energy transfer from the medium to the agglomerates. Integration of this equation within the limits $E = 0$ and E yields Equation 6.65:

Equation 6.65:

$$\frac{(N_n)_E}{(N_n)_{E=0}} = e^{-a\frac{E}{V\cdot\sigma}}$$

The number of the dispersed agglomerates $(N_d)_E$ equals the difference between the number of the agglomerates initially present and the number of agglomerates still present after the energy has acted.

Equation 6.66:

$$(N_d)_E = (N_n)_{E=0} - (N_n)_E$$

Combination of Equations 6.65 and 6.66 yields the ratio of the number of dispersed agglomerates to the number of agglomerates that have not been submitted to energy action.

Equation 6.67:

$$\frac{(N_d)_E}{(N_n)_{E=0}} = 1-e^{-a\frac{E}{V_T\cdot\sigma}} = P_E$$

Equation 6.67 is the mathematical expression of the energetic partial probability P_E. The energetic or breaking probability depends upon the dissipated mechanical power when the dispersion reaches a steady state, that is to say when all the temperatures are steady and do not change. If, for example, the dispersion is carried out at a mechanical power of 2000 Watts, then the relevant energy E would be 2000 Joules.

Total probability P_T

Since both probabilities must occur simultaneously for agglomerates to be dispersed, the total probability, defined as the ratio of the number of dispersed agglomerates N_d to the number of agglomerates initially present N_0, equals the product of P_g by P_E, i.e.

Equation 6.68:

$$P_T = \frac{N_d}{N_0} = \left(1-e^{-k\cdot\frac{V_e}{V_T}\cdot t}\right) \times \left(1-e^{-a\cdot\frac{E}{V_T\cdot\sigma}}\right)$$

The total probability P_T is therefore the product of two partial probabilities, whose values may vary between 0 and 1. It states that the efficiency of the dispersion process will be higher if the exponents in the exponential terms increase.

6.2.2 Practical conclusions from the probability relationship for the dispersion process

Some important conclusions for production practice may be drawn from the probability relationship describing the dispersion operation. The constant k is determined by the medium viscosity, because the transport speed of the agglomerates into the effective volume elements is viscosity-dependent. However, a reduction in viscosity may possibly give rise to counteracting effects, for example a decrease in mechanical stirring performance, leading to poorer transfer. As explained in Section 6.1, viscosity changes may simultaneously exert a positive and a negative effect. Thus the direction in which k is influenced depends on whether the positive effects from viscosity increase or viscosity decrease gain the upper hand.

Both the mode of operation of the milling machine and the quality of medium for transferring forces are reflected in the constant a, the energetic efficiency factor. Hence, a is also viscosity-dependent.

The ratio of the effective volume to total volume (V_e/V_T) can be regulated by the volume of the grinding media in attritors, bead and sand mills. On the other hand, long wetting-out and thorough pre-mixing at raised temperature lead to a reduction in the mean binding energy σ.

The grinding time can be chosen as desired; nevertheless, when the energetic factor a in the second exponential term becomes decisive for the dispersion process, prolonging the grinding time beyond a certain limit cannot raise the efficiency further. Accordingly, Equation 6.68 takes into consideration that insufficient mechanical performance cannot be made up by higher energy input and vice-versa.

The energy density E/V_T is determined by the mechanical factors of the dispersing machine. It can be controlled within certain limits by the stirring speed in the case of bead and sand mills, and by the applied pressure in roll mills. The energy density must always be sufficient to overcome the cohesion forces in the agglomerates as otherwise the dispersion effect will also not be increased a by longer milling time.

An important result of the theory is that the success of a dispersion can be limited either by its duration or by the mechanical power input and that longer milling time can only to a certain extent compensate a shortage of mechanical power and vica versa.

Examples for quantitative application of the probability relationship are already available [159–161].

6.3 Dispersibility of pigments

6.3.1 Critical approach to the definition of pigment dispersibility according to DIN ISO 18451-1

In accordance with the German standard DIN ISO 18451-1, the dispersibility of a powdered material is defined as the property of this material to be dispersed. The standard further states that the dispersibility depends on the **wettability** and on the **strength of the contact sites** between the structural units of the agglomerates.

Although this definition is not wrong, it does, however, not bear a critical examination due to its oversimplified interpretation of considerably more complicated facts. On discussing "ease of dispersion" (see Section 4.2.1.2) it was mentioned that this term, defined as a characteristic quantity for the dispersibility, was not a typical pigment property, but that it also depended on the dispersing conditions. Figure 6.1 lists the factors influencing the dispersibility of pigments.

It is easy to see that a definition of the dispersibility of a pigment is actually an attribute denoting the efficiency of the whole system, thus including pigment, medium, dispersing machine and other conditions. Hence, it has little value to characterise dispersibility, in whatever manner, without quoting the dispersing conditions.

6.3.2 Testing and stating the dispersibility as specified in DIN EN ISO 8780 and DIN EN ISO 8781

The pigment processor is better helped with the definition of **ease of dispersion** given in the standard DIN EN ISO 8780-1, as quoted in Section 4.2.1.2 on page

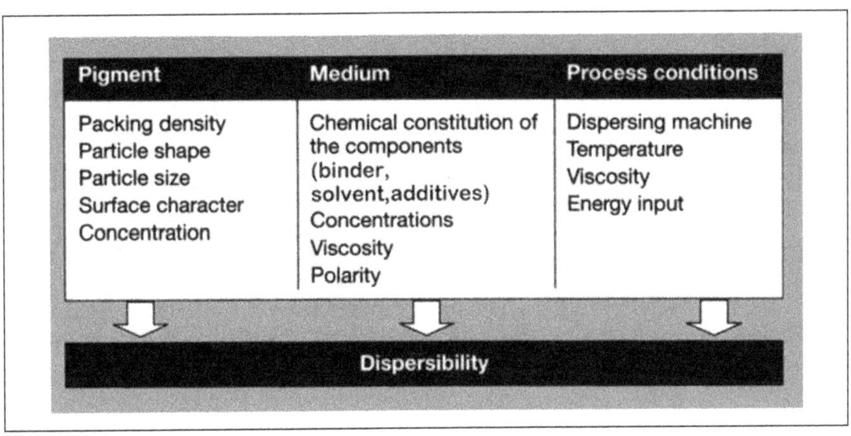

Figure 6.1: Factors influencing pigment dispersibility

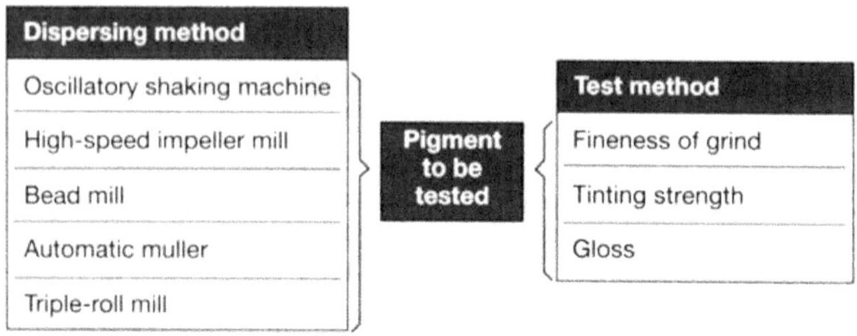

Figure 6.2: Possible combinations of dispersing method and testing criteria in accordance with the modular principle (DIN EN ISO 8780 and DIN EN ISO 8781)

123. This definition is more in line with the facts governing pigment dispersion. Section 6.3.4 deals with another definition also leading to an appropriate quantity for pigment dispersibility.

The pigment technologist is interested in the dispersibility of a given pigment under his own working conditions, and not in how well or how badly the pigment can be dispersed under the test conditions of his supplier. From this standpoint, the standards committee for Pigments and Extenders (NPF) of the DIN German Institute for Standardization e.V. compiled a procedure for testing and stating the dispersibility under selected conditions, i.e. it enables an assessment of dispersibility in a variety of binder systems and milling devices according to different testing criteria, and it is possible to combine these three conditions according to the modular principle. Such a test method was specified in the former DIN 53238 series of standards and later adopted by ISO (the International Organization for Standardization) as ISO 8780 and ISO 8781. In the meantime, these ISO standards are implemented as European Standards EN ISO 8780 and EN ISO 8781, and are mandatory for all member countries of CEN, the European Organization for standardization. All European Standards are implemented by the CEN members, so in Germany as DIN EN ISO 8780 and DIN EN ISO 8781. The possible combinations are illustrated schematically in Figure 6.2. Which test method should be applied, can be freely negotiated between the interested parties, for example supplier and pigment processor. The binder systems to be used are not specified, due to the multiplicity of systems encountered in practice. Nevertheless, advice of a general nature is given in the different parts of DIN EN ISO 8780.

Equation 5.54, derived according to *von Pigenot's* method for calculating the dispersion expenditure required for the development of tinting strength, was slightly modified in the former standard DIN 53238-22. Two dispersion stages were chosen, of which stage 1 corresponded to a relatively short milling time just sufficient to ensure complete pigment homogenisation in the medium. Stage 2 was laid out in such a manner that the final tinting strength is nearly attained. The dispersion expenditure required for stage 1 was attributed the nominal value 100. Thus, the dispersion expenditure for stage 2, i.e. the actual dispersion hardness *DH*, automatically resulted in a percentage of the first value above 100. Hence, the equation for *DH* became:

Equation 6.69:

$$DH = 100\left(\frac{F_2}{F_1} - 1\right)$$

where

DH = dispersion hardness (expression no longer used for binder systems)
F_1 = tinting strength at stage 1
F_2 = tinting strength at stage 2

For the sake of better international understanding, dispersibility in DIN EN ISO 8781-1 is given in terms of **increase in tinting strength** (in per cent), whereby the numerical value of the latter is identical with that of the former dispersion hardness. Thus, IS (for increase in tinting strength) appears in place of DH in the pertinent equation.

It should be noted that the numerical value of the increase in tinting strength depends on the principle for selection of the dispersion stage.Thus, only those *DH*-values based on equivalent dispersion stages can actually be compared. Moreover, when choosing a suitable dispersion stage, both prerequisites mentioned must be fulfilled, i.e. F_2 should almost correspond to the final tinting strength and it should be substantially higher than F_1, as the simplifications applied on deriving Equation 6.69 are only justified under these conditions.

Since two stages are not sufficient for characterising the dispersion behaviour of new pigment types, a series of stages – preferably four to eight – has been proposed to determine the tinting strength development curve in these cases (see Section 5.3.1).

6.3.3 Assessment of the dispersibility by means of a half-life value

D. von Pigenot suggested rating the dispersibility by means of a more demonstrative characteristic, namely the **half-life value**, instead of E_0 (the expenditure in the

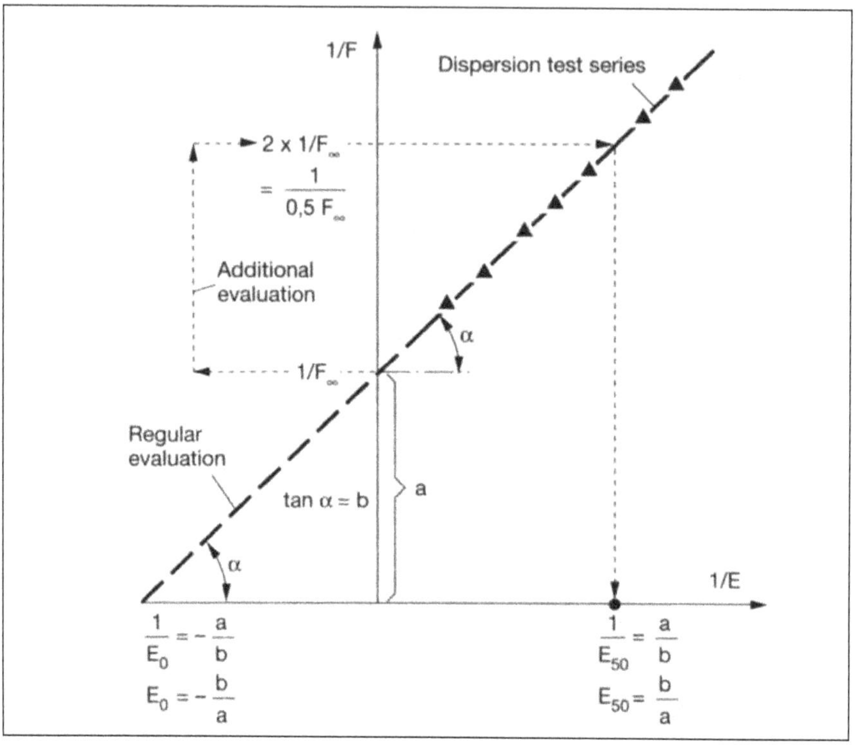

Figure 6.3: Interpretation of the half-life value as a measure for the dispersibility. F tinting strength: F_∞ final tinting strength: E energy input: l/E_0 intersection with the abscissa: a intersection with the ordinate: b slope of the straight line: E, half-life value of the energy input

dispersion process as expressed in Equation 5.54), or of *DH*. The half-life value is defined as the work[1] of dispersion required to develop 50 % of the final tinting strength.

In the scheme shown in Figure 6.3 it may easily be verified that simple geometric similarities result in the half-life value E_{50} being numerically equivalent to E_0, but with a positive sign [136]. Considered formally, it is the energy input, i.e. the amount of energy required to obtain half of the final tinting strength and, strictly speaking, it should be expressed in the usual SI – units of energy, i.e. kWh, Ws or J. Nevertheless, as the energy input is directly proportional to either milling time, in the case of ball, bead or sand mills, or to the number of passes at constant

[1] Since work is defined as the product between power and dispersion time, the expression 'work' may be somewhat misleading. In this only the time is changed whereas the mechanical power is considered to be constant, yet at an unknown value.

pressure through a triple-roll mill, or to the number of revolutions when using an automatic muller, it can also be indirectly expressed as minutes, hours, number of passes or revolutions, depending on the type of milling device employed.

6.3.4 Pragmatic definition for pigment dispersibility

For a better understanding of another concept for pigment dispersibility, we shall resort, initially, to an example from practice: Two different commercial grades of nickel titanium yellow, denoted A and B, had to be compared with regard to their dispersion behaviour. The samples were dispersed in the same medium by means of an oscillatory shaking machine. The criterion for the comparison was the milling time in minutes required to obtain a fineness of grind of 10 µm. This value was determined by graphical interpolation with a linear representation of the dispersion curve (double log graph). Thus, it was determined that sample A required 16.2 and sample B 17.8 minutes millig time, and A was correspondingly evaluated as the grade showing better dispersibility under the given conditions. So far our example.

This rating would have been correct, if the pigment content in both mill bases had been the same. However, this was not the case. On preparing the mill base with sample A, a suitable consistency was obtained with a pigment concentration of 30 %, whereas with sample B, owing to its lower binder demand, a concentration of 35 % was required to obtain the same consistency. It was considered reasonable to make allowance for the differing binder demands of both grades when preparing the mill bases, as pigment B would otherwise have been handicapped due to the lower mill base consistency.

Thus, when facts are studied more closely, about 10 % more energy input was necessary to reduce the oversize of sample B to a level of 10 µm. However, about 17 % more pigment was dispersed to this degree of quality than in the case of sample A. Is pigment B even now, under these conditions, less easy to disperse than pigment B? In order to be certain, it is best to calculate the ratios between the amount of pigment and the corresponding energy input. For the sake of simplicity, let the prepared amounts of mill base be 100 g. Therefore, the test with grade A contains 30 g and the one with grade B 35 g of pigment. Dividing the pigment amounts by the corresponding milling times yields:

$$\frac{30}{16.2} = 1.851 \text{ g/min} \quad \text{for pigment A}$$

and

$$\frac{35}{17.8} = 1.966 \text{ g/min} \quad \text{for pigment B}$$

Hence, it follows that 6.2 % more pigment B was dispersed per milling minute than pigment A. The dispersion expenditure was thus more effective in the case of pigment B and, consequently, B is actually the better dispersible grade.

We now have a new definition for the dispersibility, although it was introduced originally in a paper published as far back as 1973 [162], where it was pointed out that not only the dispersion expenditure, but also the dispersed amount of pigment has to be taken into account for a proper rating of pigment dispersibility. *Eckhoff* later generalized this concept by applying it to all powdered materials and proposed that their dispersibility (D) by best characterised be the ratio of the amount of powder (P) dispersed to the required energy input (E) [163], i.e.:

Equation 6.70:

$$D = \mathrm{k} \frac{P}{E}$$

The dispersibility has therefore the dimension mass per unit energy, and is thus a measure for the amount of pigment that can be dispersed in a given medium under specific conditions at the expense of one unit of energy. It is possible to express the amount of pigment in terms of mass as well as volume concentration. For the energy input data, use may be made of the alternatives mentioned in Section 6.3.3. (i.e. indirectly by quoting milling time, number of passes, etc.). Since the proportionality constant k is system-specific, it can be disregarded when the dispersion behaviour of pigments is tested in one and the same binder/solvent system.

We may conclude from these statements that the assessment of pigment dispersion behaviour as specified in DIN EN ISO 8780 and DIN EN ISO 8781 is, strictly speaking, only valid when the pigments to be compared are identical or differ only insignificantly as to their binder demand. This is usually the case when controlling production batches of a pigment with regard to their conformity with the standard or the reference sample. When comparing dispersibility using the half-life value as a criterion, this dimension should also be stated in reference to the dispersed amount of pigment if the pigment concentrations in the tests differ.

As the binder demand of a pigment must be taken into consideration when assessing its dispersibility, we may also infer that each medium exhibits a specific pigment concentration which makes best allowance for the binder demand and the flow properties of the pigment with regard to its dispersion behaviour.

The dispersibility value obtained at this optimum pigment concentration – with respect to the dispersion process –, and none other, is the required dispersibility of the pigment in the given medium under the chosen milling conditions.

Table 6.12: Example 5: Pigment concentrations of the eight trials, the corresponding milling times required for a fineness of grind of 10 µm and the resulting dispersibilities

Trial	Pigment concentration [g/100 g]	Milling time [h]	Dispersibility [g/h]
1	6	16.00	0.375
2	9	16.50	0.545
3	12	17.75	0.676
4	14	19.00	0.737
5	16	21.50	0.744
6	18	25.00	0.720
7	20	32.00	0.625
8	23	58.00	0.397

By assessing the dispersibility values for a series of pigment/medium combinations, it is possible to determine the optimum pigment concentration as, logically, this particular concentration must yield the best dispersibility value. This will now be demonstrated by a practical example.

Example 5

In order to determine the dispersibility (D) of a BON, Mn-lake (C.I. Pigment Red 52:2, N° 15860), mill bases were formulated with increasing pigment content in an 30 %w solids alkyd resin in white spirit, and dispersed in laboratory ball mills. In all tests, the milling time required to attain a fineness of grind of 10 µm was determined. The corresponding **pigment concentration : milling time** ratios were then calculated. The results are given in Table 6.12.

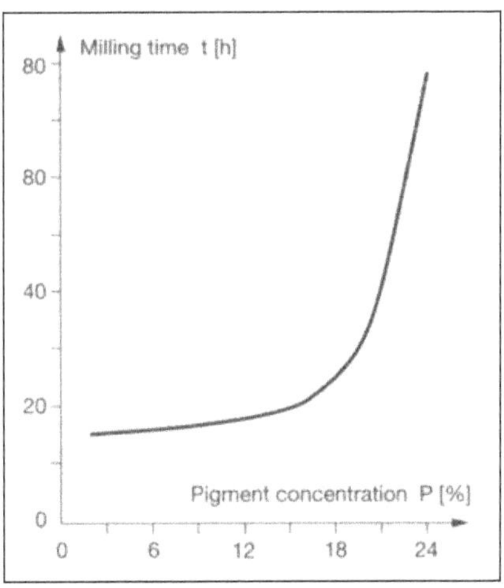

Figure 6.4: Increase in the dispersing time required to attain a fineness of grind of 10 pm for an organic pigment (Pigment Red 52:2) in dependence on pigment concentration in an alkyd resin/white spirit paint system

Obviously, the highest D-value (0.744) is obtained at 16 % pigment concentration. Thus, as a first approximation, 16 % is the optimum pigment concentration and 0.744 g/h the dispersibility of the tested pigment under the given dispersion conditions.

The milling time is plotted as a function of the pigment concentration In Figure 6.4. The curve corresponds to a monotonic increasing exponential function. Denoting the pigment concentration as P and the milling time as t, a computer regression analysis of the data pairs indicated the following relationship to be the best fit:

Equation 6.71:

$$t = a + b\,e^{cP}$$

with the following values for the parameters a, b und c:

$$a = 15.98 \qquad b = 0.036 \qquad c = 0.31$$

According to Equation 6.70 D is equal to P/t (t corresponds to E; the medium-dependent constant k can be disregarded, as the medium is the same for the whole series of tests). Hence, by inserting the term for t from Equation 6.71 in Equation 6.70), we obtain:

Equation 6.72:

$$D = \frac{P}{15.98 + 0.036\,e^{0.31P}}$$

The curve of D as a function of P calculated according to Equation 6.72 (i.e. solid curve in Figure 6.5) corresponds to a growth function with non-constant growth factor in the segment examined.

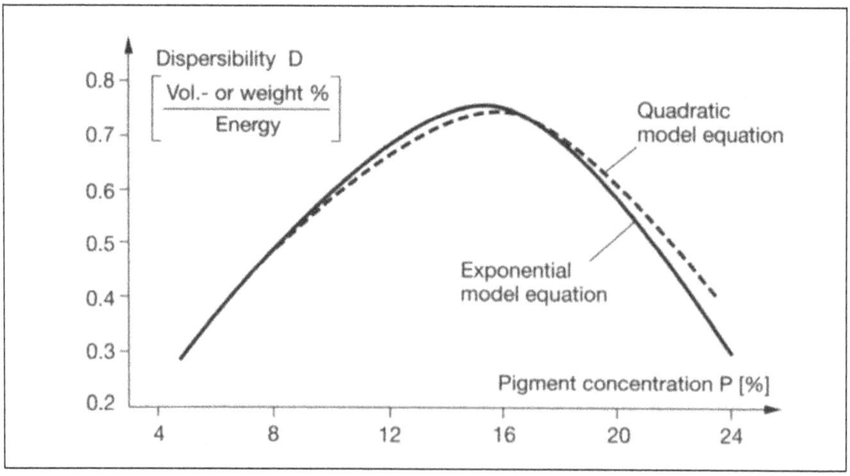

Figure 6.5: Dispersibility as a function of pigment (weight) concentration according to exponential (solid line, and quadratic model equation (dashed line)

As the pigment concentration P appears in both numerator and denominator in Equation 6.72, we can recognize its ambivalent effect on dispersibility. The positive effect outweighs at low concentrations, but from about 12 % upwards, due to re-agglomeration and increase in viscosity, the P in the denominator starts taking stronger effect, so that the increase slows and becomes retrograde from a specific concentration. The maximum at the turning point corresponds to the dispersibility of the pigment in question, and the relevant concentration, which may be termed $P_{op,}$ is the optimum pigment concentration in the given binder/solvent system.

Empirical assessment of D_{max}

For an exact determination of the maximum, Equation 6.72 should normally be differentiated with respect to P, the derivative set equal to 0, and P_{op} calculated from this new equation as a function of the parameters. D_{max} can then be calculated by inserting the value obtained for P_{op} in Equation 6.72. However, due to the exponential term, it is not possible to obtain an explicit expression for P_{op} by simple algebraic means. The task becomes far easier if the upper part of the curve is considered to be a parabola, i.e. we describe the zones on the left and on the right of the peak of $D = f(P)$ by a second degree equation. The relationship between D and P then results in the expression:

Equation 6.73:

$$D = a P^2 + b P + c$$

Assessment of the parameters a, b und c requires three equations. Taking the P-values giving the peak, and their corresponding D-values, from Table 6.12, we may thus draw up the following system of equations:

1. $0.676 = 144 a + 12 b + c$
2. $0.744 = 256 a + 16 b + c$
3. $0.625 = 400 a + 20 b + c$

Solving this system of equations gives the following values for the parameters:

$$a = -0.00584$$
$$b = 0.18063$$
$$c = -0.65000$$

Consequently, for the case we are considering Equation 6.73 is:

Equation 6.74:

$$D = -0.00584 P^2 + 0.18063 P - 0.65000$$

Table 6.13: Empirically determined dispersibility values and equivalent optimum concentrations for several organic and inorganic pigments

Pigment	Optimum concentration [%]	Dispersibility D_{max} [g/h]
BONS, Mn-lake (C.I. P.R. 52:2)	15.5	0.75
Cu-phthalocyanine blue (easy dispersible)	16.0	0.82
Toluidine red	26.1	0.92
Prussian blue	21.3	1.05
Lead chromate	36.0	1.65
BONS, Mn-lake (C.I. P.R. 48:4)	15.1	1.69
Lead chromate/molybdate	36.5	1.72
Nickel titanium yellow	36.6	2.01
Lead chromate (easy dispersible)	46.0	3.42
Titanium dioxide	50.0	3.60

By differentiating Equation 6.74 with respect to P and equating the derivative to 0, we obtain for P_{op}:

$$P_{op} = \frac{b}{2a} = \frac{0.18063}{0.01168} = 15.46 \approx 15.5$$

To determine the required dispersibility D_{max}, the value for P_{op} is inserted in Equation 6.74, and thus we obtain:

$$D_{max} = 0.747 \approx 0.75$$

The broken line in Figure 6.5 shows the D = f(P)-curve according to Equation 6.74. It can be clearly seen that the deviation is insignificant when compared to the curve according to Equation 6.72. Calculation of D_{max} according to Equation 6.72, using a value of 15.5 for P, gives 0.76 – a deviation of 1.3 %.

Table 6.13 lists the P_{op} and D_{max} values for several organic and inorganic pigments obtained using the method explained in Example 5.

Thus, it is possible to assess the dispersibility and optimum pigment concentration in a given medium on the basis of three experiments. However, sufficient accuracy is only obtainable when the concentration range for the experiments is chosen to include the peak of the parabola, i.e. the optimum concentration, as was the case in Example 5. Further examples can be found in the literature [162, 164]. In place of the fineness of grind, other properties of the pigmented system, such as tinting strength, lightness flop or gloss, may be used as criteria for the dispersibility [147, 173].

6.4 Methods for mill base optimisation

In accordance with the explanations in Section 6.1, mill base optimisation in connection with pigment processing is understood to be the quantitative correlation of the mill base ingredients to obtain the best result under the given dispersing conditions. 'Best result' is the achievement of the desired quality level at the lowest possible expenditure As the application properties of pigments, such as gloss, tinting strength, hiding power, fineness of grind or transparency, develop differently in each case during dispersion, the choice of the quality criterion is decided firstly by the kind of pigment (organic or inorganic, hiding, transparent or semi-transparent etc.), and secondly by the most important property of the final product.

It is obvious that the economics of a milling operation will be better, the higher the pigment content in the mill base. For this reason, pigment processors generally strive to disperse a high loading of pigment in a binder solution having the lowest possible solid content. *F. K. Daniel* pointed out these relationships as far back as 1946, and proposed a method to determine optimum mill base compositions for ball mills [165].

6.4.1 Daniel flow-point procedure

The flow-point procedure according to *Daniel* comprises pugging a given pigment mass by titration with a binder solution and whilst vigorously kneading it with a spatula. This is continued until the paste flows as a continuous thread from the blade, and this operation is carried out succesively with a series of binder solutions of increasing concentrations. By plotting the volume of binder solution required against the corresponding binder concentrations in a Cartesian diagram, a flow-point curve is obtained which exhibits a minimum. The binder concentration at this point is considered to be the optimum. The appropriate pigment concentration results from the amount of pigment used. Accordingly, a binder solution of about 15 % solid content should be chosen in the example illustrated (Figure 6.6).

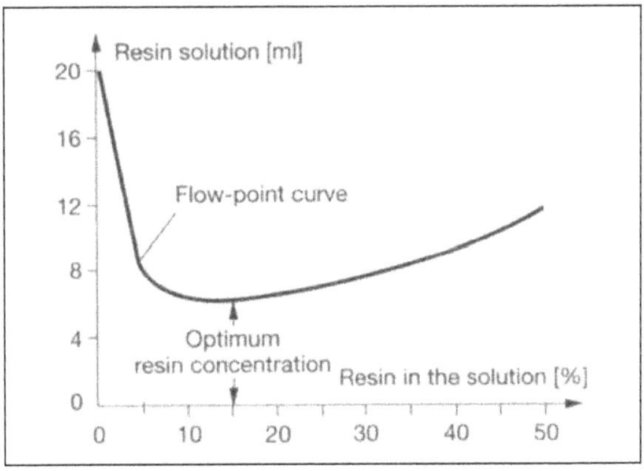

Figure 6.6: Example for the determination of the optimum mill base formulation according to Daniel's flow-point method

In Daniel's opinion, ball mill base compositions assessed by using the flow-point method apply equally well to stirrer-type ball mill grinding, e.g. attritors, bead mills.

For high-speed disc impellers Daniel recommends the formulation resulting from the **wet-point**. This is an earlier stage of the flow-point determination and is reached when the mixture of pigment and binder solution just becomes a coherent, homogeneous mass that starts to smear. The procedure used to calculate the formulation is the same as before, but using the wet point instead of the flow-point.

Optimum mill base formulations based on the Daniel flow-point method may also be assessed by the aid of a triangular co-ordinate system. The percentages of the components are entered into a concentration triangle, and the appropriate concentrations taken from the point of maximum pigment content. The procedure has been described in detail by *Rohrer* [166].

Using the concentration triangle, *Göring* and *Paksoy* were able to demonstrate that optimum mill base compositions for the sand mill may be determined by carrying out laboratory tests in a vibrating ball mill, also known as **high-speed laboratory mill** or just **quickie mill**. The results of a series of tests with varying binder and pigment concentrations performed parallel in a vibrating ball mill and in a sand mill, lead to the conclusion that the best efficiency is obtained for both milling devices using formulations within a narrow zone of the concentration triangle [167].

Research performed by *Liehr* showed that the dispersion carried out with formulations determined according to the flow-point method can be improved in many cases by a slight addition of binder solution. This was found to be the case with formulations for ball mills as well as for attritors, sand and bead mills. Conversely, in the case of triple-roll mills the best result was obtained by direct use of the mill base composition derived from the wet-point determination [168].

Comment upon the Daniel flow-point method

Although the Daniel flow-point method has proven useful in production, it has some drawbacks that cannot be overlooked.

1. The estimation of the flow-point is submitted to the subjective appraisal of the tester.
2. The shearing forces exerted during the test are insufficient for thorough dispersion of the pigment. In most cases, this results in too low a value for the assessed binder demand .
3. Anomalous flow properties, possibly arising from pigment/binder interactions may falsify the flow-point.

Taking these drawbacks into consideration, we may conclude that the Daniel flow-point is too uncertain in the case of organic pigments due to their high specific surface area and complex rheological behaviour. The use of the flow-point method thus remains largely confined to inorganic pigments.

6.4.2 Guggenheim factor

High-speed impeller mixers, e.g. of the dissolver type, have become indispensable for pigment processing in solvent-based systems. Their dispersing effect results from the extreme shear forces caused by laminar streaming of the relatively high-viscosity mill base. However, optimum dispersion is only achieved, when certain prerequisites regarding the dimensions of the mixer, its operation and the amount and consistency of the mill base are fulfilled. Thus, specific interrelations between tank and impeller diameter, depth of the mill base and distance between impeller and bottom of the tank have to be taken into account. In addition, mill base consistency and peripheral speed of the impeller should only be allowed to vary within a narrow, fixed range. If all these conditions are met, a defined flow pattern occurs that is known as **doughnut effect** and is characterised by a flow movement of the whole of the mill base from the wall to the impeller on a nearly radial course (formation of a torus), without any dead zones or turbulences. This absolutely laminar flow ensures the highest shearing effect. The apparatus, dimensional relationships, and mode of operation have been discussed repeatedly in the pertinent literature [165, 168–171 and DIN EN ISO 8780-3].

For the assessment of suitable mill base compositions for high-speed impeller mixers, *Guggenheim* (1958) proposed an empirically developed expression which, to be accurate, represents a correction of the Daniel wet-point [172]. Guggenheim studied the relationships between viscosity and solid content of the binder solution, and the binder demand of the pigment in a series of systems containing oil-modified alkyd resins, and succeeded in determining the following equation that allows the calculation of a correction factor:

Equation 6.75:

$$\frac{F}{C} = 90 + \frac{FK}{145} + \frac{\eta}{4}$$

where

F/C = correction factor, the Guggenheim factor,
FK = solid content of the binder solution in percent,
η = viscosity of the binder solution in Pa·s.

According to Guggenheim, the amount of binder solution required for dispersion in a high-speed impeller mixer is obtained by multiplying the **oil absorption value** of the pigment by the Guggenheim factor. The oil absorption value of a pigment, a quantity characterising its binder demand, is the number of millilitres of refined linseed oil that must be added to 100 g of pigment to reach the wet point. The determination is specified in DIN EN ISO 787-5 and in ASTM D 1483.

Liehr introduced a slightly modified version of the Guggenheim method. He replaced the oil absorption value by the amount of binder solution required to reach the wet point. In all other respects the procedure is the same as that of Guggenheim [168].

The dissolver can be used for direct determination of a suitable mill base composition in the case of laboratory tests. The binder solution is first charged into the tank, and the pigment is then gradually added with the disc rotating initially at low speed. The increase in viscosity is kept within limits by the increasing temperature. The appropriate pigment concentration is reached, when a correct doughnut effect is exhibited at a peripheral speed of 20 to 25 m/s and a temperature of about 70 °C. The solid content of the binder solution depends on the type of binder and whether the operation is intended only as a pre-mix or as a proper dispersion. As a rule of thumb, about 55 to 70 % solid content is recommended for high-molecular binders, and 65 to 90 % for medium-molecular types.

6.4.3 Efficiency of the dispersion process: new criterion for optimizing mill bases

As mentioned in Section 6.1, the dispersion process depends not only upon the nature and concentration of the pigment, but also on further factors affecting other components of the mill base, their concentration, physical qualities such as temperature and viscosity, and the dispersing machine itself. Hence, it follows that pigment dispersibility in the sense of the definition proposed in Section 6.3.4, i.e. the ratio between mass of pigment dispersed to a specified quality level and the relevant energy input required, is a measure for the efficiency of the whole dispersion system.

Thus, dispersibility could be used as a criterion for optimizing mill base compositions. The optimum formulation is the one which yields the highest value for the dispersibility. These facts were pointed out in several papers published in the past, and the term **degree of efficiency** (German: Nutzgrad) was proposed for this new criterion. The concept 'degree of efficiency' is thus nothing other than the dispersibility acting as a criterion for the mill base optimisation [162, 164]. Further papers demonstrated the use of various performance properties, such as fineness of grind, tinting strength or lightness flop, to assess the degree of efficiency [173, 174].

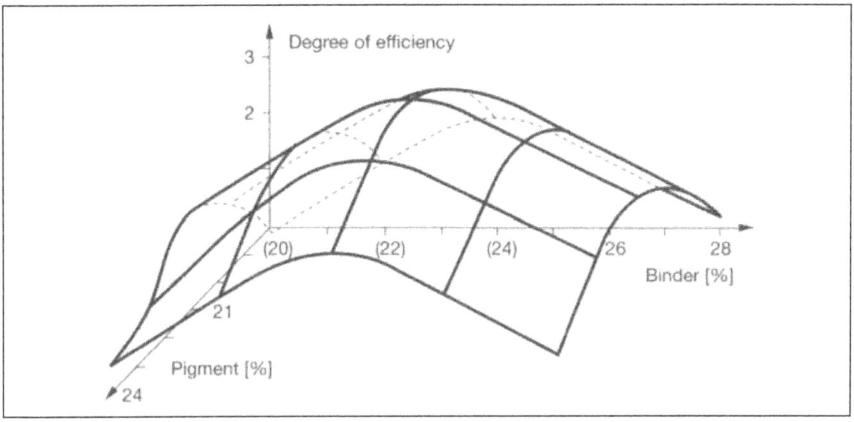

Figure 6.7: Degree of efficiency of a dispersion process as a function of pigment and binder concentration in the mill base

Considering the discussions in the preceding sections on wetting, viscosity and stabilisation, it is apparent that the effects of binder concentration will parallel the effects determined for pigment concentration, i.e. an increasing concentration in the mill base has initially a positive and later a negative effect on the dispersion process. The binder provides good transfer of shear forces and stabilisation of the broken-down pigment particles. However, too high a concentration of binder is nevertheless detrimental, as it causes an increase in viscosity, and this results in inhibition of shear and impact forces. It also results, according to Washburn's equation, in the retardation of wetting, so that negative effects will predominate when a specific binder concentration.is exceeded.

Figure 6.7 illustrates the dependence of the degree of efficiency upon pigment and binder concentration for a phthalocyanine blue (C.I. Pigment Blue 15:1; N° 74160) in a waterborne paint system. It can clearly be seen that the degree of efficiency passes through a maximum with both increasing pigment and binder content. If we wish to optimize the composition of a given pigment/binder/solvent system, we will have to determine those concentrations of the three components which yield the highest value for the degree of efficiency, or, expressed in more general terms, the aim is to find those levels of the factors which give an optimum response. As we are dealing with a ternary system, it is sufficient to determine the optimum concentrations of pigment and binder – the solvent concentration will result from the difference of the sum of the other two to 100 %.

Empirical optimisation of P and B

The simplest function yielding a maximum is a polynomial of degree two. It was shown in Section 6.3.4 that the dependence of dispersibility upon pigment concentration can be well represented by a quadratic equation, and thus a quadratic equation is selected as a model for the dependence of the degree of efficiency on pigment and binder concentration. Using the symbols P and B for pigment and binder concentration respectively, and N for the degree of efficiency, the model equation takes the form:

Equation 6.76:

$$N = k_0 + k_1 P + k_2 P^2 + k_3 PB + k_4 B + k_5 B^2$$

This model takes into consideration all factors up to the second degree, including a possible interaction between both variables. It results in a parabolic surface and thus has the advantage that the peak can easily be found. By partial differentiation of Equation 6.76 with respect to P and to B, and by setting the derivatives equal to zero, we obtain two equations in which the values we require P_{op} and B_{op} are the two unknowns:

$$\left(\frac{\partial N}{\partial P}\right)_B = k_1 + 2\,k_2\,P_{op} + k_3\,B_{op} = 0$$

$$\left(\frac{\partial N}{\partial B}\right)_P = k_4 + 2\,k_5\,B_{op} + k_3\,P_{op} = 0$$

The solution of these equations results in:

$$P_{op} = \frac{k_3 \cdot k_4 - 2k_1 \cdot k_5}{4k_2 \cdot k_5 - k_3^2}$$

$$B_{op} = \frac{k_1 \cdot k_3 - 2\,k_2 \cdot k_4}{4k_2 \cdot k_5 - k_3^2}$$

The parameters k_0, k_1 etc. are the polynomial coefficients. They can be estimated by performing a series of at least six trials with different pigment and binder concentrations and evaluating the result using the **method of least squares**. However, handling six equations is a fairly protracted affair if an electronic computer is not available. Moreover, the parameters cannot be estimated independently from one another. It is thus advisable to carry out their determination by means of experiments based on **statistical design**.

Special experimental plans have been compiled in which the coefficients are orthogonally related with each other, and thus their determination can be per-

Table 6.14: Experimental plan with corresponding results (degree of efficiency N) for nine trials with two factors at three levels for the determination of the optimum mill base formulation for a phthalocyanine blue pigment in an aqueous paint system

Trial	Pigment in millbase [%]	Binder in millbase [%]	Result [Efficiency degree N]
1	18.0	20.0	0.575
2	18.0	24.0	1.320
3	18.0	28.0	0.954
4	21.0	20.0	1.359
5	21.0	24.0	1.879
6	21.0	28.0	1.462
7	24.0	20.0	1.151
8	24.0	24.0	2.155
9	24.0	28.0	1.398

formed by simple linear relations. They are termed **Orthogonal Composite Designs**, and imply several experiments with the factor levels chosen so as to surround a central experiment.

A further advantage of statistical experimental designs arises from the possibility to determine whether individual terms in the model can be disregarded because their coefficients may not be significant enough for the course of the target function. The response surface shown in Figure 6.7 was obtained using such an experimental design. The design matrix itself, involving nine trials with just as many combinations of the varying factor levels, and the resulting degrees of efficiency obtained are given in Table 6.14. The statistical evaluation resulted in 21.3 % and 24.1 % as the optimum levels for pigment and binder concentration respectively. These values, inserted in Equation 6.76, gave a maximum degree of efficiency of 3.37. The interested reader will find a detailed description of statistical experimental design and its realisation together with examples for its application in a paper already quoted [174].

A quantitative optimisation of the mill base components by using the tinting strength as the target function is also possible as well, particularly in the case of organic pigments. In the same manner as the degree of efficiency, the tinting-strength curve for a given amount of pigment also shows a maximum if pigment or binder content in the mill base are increased. Figure 6.8 illustrates the dependence of tinting strength on pigment concentration and milling time for the same pigment and binder system as in Figure 6.7.

An example for a mill base optimisation by statistical experimental design using the tinting strength as target function can be seen in Figure 6.9, where the **response surface** for a perylene maroon (C.I. Pigment Red 179, N° 71130) is illustrated.

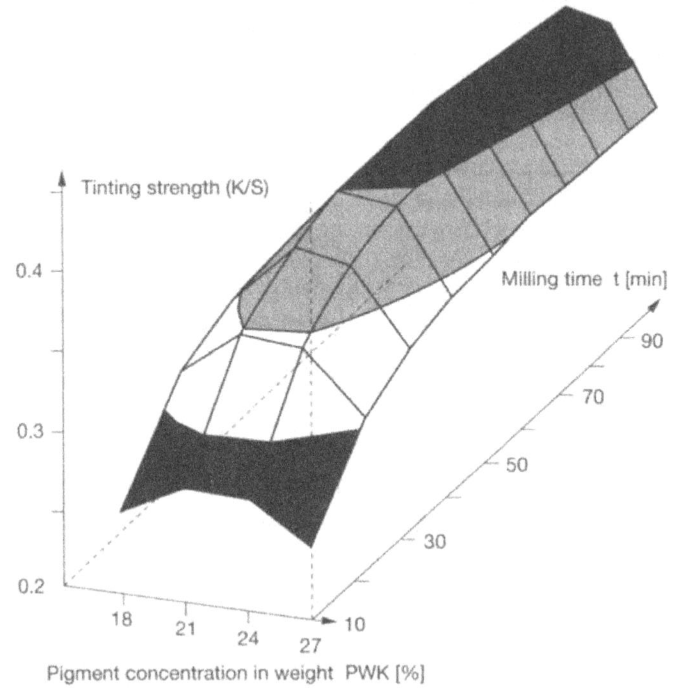

Figure 6.8: Tinting strength development on dispersing in dependence upon milling time and pigment (weight) concentration in the mill base

The medium used for the series of trials shown in Figure 6.7 was also used in this case. The values 20.3 % and 17.8 % were found for optimum concentrations of pigment and binder respectively for this perylene maroon grade.

After the optimum concentrations of pigment, binder and solvent in the mill base have been determined, it is still necessary to discover whether the pigment paste produced under these conditions is sufficiently stable. In other words, one must not lose sight of the third and most important stage of the dispersion process, namely stabilisation. The degree of efficiency also expresses the tendency to re-agglomeration, which must necessarily be weakest in the system with the maximum degree of efficiency. However, it would be wrong to conclude that the system must thus have a good storage stability, and the optimised paste must be submitted to a subsequent stability test. In some cases, if the result is negative, a slight deviation from the optimum concentrations towards a lower pigment to binder ratio will be necessary to ensure that enough binder molecules are available during the dispersion process to sufficiently stabilise the dispersed pigment particles.

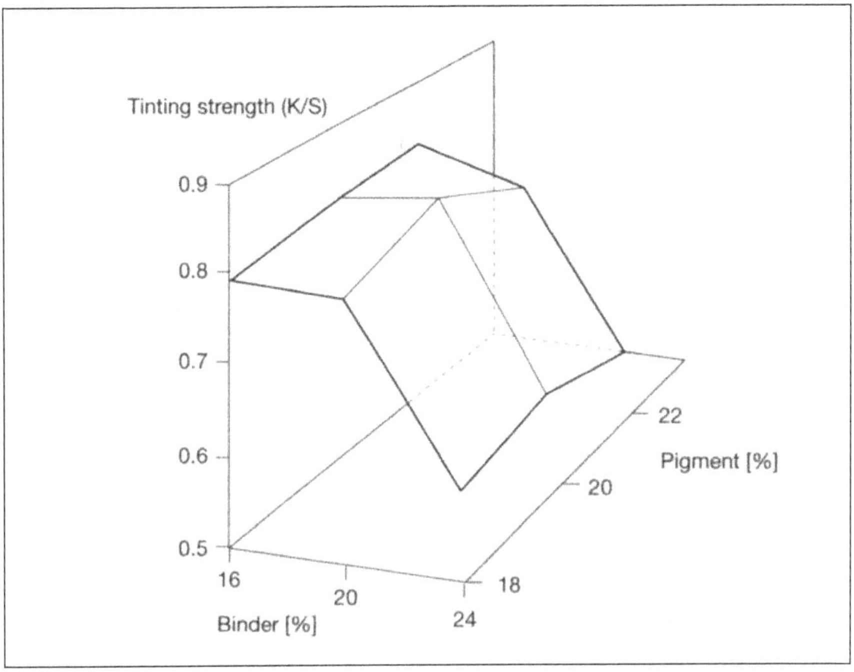

Figure 6.9: Tinting strength development in dependence on pigment and binder concentration in the mill base

Similarly, in some cases with surface-treated pigments it may occasionally be advisable to increase the binder content in the mill base if the optimum polymer concentration could result in quality problems with the coating and in a reduction of its stabilising effect (i.e. with regard to light fastness or SO_2 resistance). This type of problem, like the storage stability, will only be apparent much later.

6.4.4 Optimisation of several factors by means of an experimental design according to Plackett and Burman

If dispersion processes depending on many influencing factors have to be investigated, it is advisable to resort to experimental designs such as those developed by *Plackett* and *Burman* [175]. Thus it is possible to examine N - 1 variables by performing N trials. The method accepts that, owing to the relatively low expenditure of work, a great deal of information will be lost. However, the Plackett/ Burman experimental designs are particularly useful for an initial assessment of which factors are significant and which are not.

A good example for the application of such an experimental design to mill base optimisation is a study by *Carr* and *Kelly*. They investigated the effect of fifteen factors on sand mill grinding in order to determine which factors were significant and which insignificant. With the exception of two factors called **dummies** which remained unchanged, each factor was tested at two levels. The system studied consisted of a β-phthalocyanine blue and a long-oil soja alkyd resin in white spirit. The criterion used for evaluating the degree of dispersion was the tinting strength, and the samples were mixed with a standard white paint based on a titanium dioxide grade to give 1:25 reductions. The factors investigated together with their low and high levels are listed in Table 6.15. Table 6.16 shows the experimental design with the factor levels and the corresponding results, i.e. the K/S-values obtained.

Statistical analysis of the results showed that the factors A, D, E, J, K and N were insignificant. The effect of the dummy factors was used to estimate the experimental error, which proved to be negligible. Concerning the factors found to be significant, the following interpretations were obtained:

B. Size of vessel – small vessels are better
C. Diameter of discs – a large diameter is better
F. Pigment concentration – high pigmentation level is better
G. Pigment/binder ratio – low pigment/binder ratio is better
H. Pre-mixing stage – more efficient pre-mixing is better
M. Type of grinding media – sand is better than glass beads

Table 6.15: Upper and lower levels of the fifteen factors in the experimental plan according to Plackett and Burman

	Factor	Level		
		high (+)		low (-)
A	Speed of discs	4500 rpm		1500 rpm
B	Size of vessel	16.1 cm		11.8 cm
C	Diameter of discs	9.0 cm		4.5 cm
D	Time of grinding	30 min		10 min
E	Number of discs	2		1
F	Pigment concentration	12.5 %		7.5 %
G	Pigment/binder ratio	1 : 2		1 : 4
H	Premixing stage	6 hours		½ hour
(I)	Size of grinding media		FIXED	
J	Grinding media charge	24 %		16 %
K	Formulation charge	24 %		16 %
L	Temperature of grinding	55 °C		20 °C
M	Type of grinding media	Sand		Glass
N	Thickness of discs	1.5 cm		0.5 cm
(O)	Order of charging vessel		FIXED	

Table 6.16: Combinations of both levels for the fifteen factors in the experimental plan according to Plackett and Burman and corresponding results (tinting strength values)

Trial									Factor level							Response
N°	A	B	C	D	E	F	G	H	(I)	J	K	L	M	N	(O)	K/S
1	+	+	+	+	-	+	-	+		-	-	+	-	-		0.593
2	-	+	+	-	-	+	-	-		+	+	+	+	-		0.640
3	+	+	+	-	+	-	+	+		-	+	-	-	-		0.454
4	-	-	+	+	+	+	-	+		+	+	-	-	+		0.930
5	-	+	-	-	-	+	+	+		-	+	-	+	+		0.564
6	-	+	+	+	+	-	+	-		+	-	-	+	-		0.473
7	+	+	-	-	+	-	-	-	F	+	+	+	-	+	F	0.417
8	+	-	-	-	+	+	+	+	I	+	-	+	+	-	I	0.624
9	+	+	-	+	-	+	+	-	X	+	-	-	-	+	X	0.493
10	-	-	-	-	-	-	-	-	E	-	-	-	-	-	E	0.356
11	+	-	+	+	-	-	+	-	D	-	+	+	+	+	D	0.701
12	-	-	+	-	-	-	+	+		+	-	+	-	+		0.487
13	+	-	+	-	+	+	-	-		-	-	-	+	+		0.760
14	+	-	-	+	-	-	-	+		+	+	-	+	-		0.760
15	-	-	-	+	+	+	+	-		-	+	+	-	-		0.435
16	-	+	-	+	+	-	-	+		-	-	+	+	+		0.481

In order to determine the order of importance of the significant factors, a second series of experiments, in which the levels of the insignificant factors were fixed and the significant factors were again tested at two levels, was designed and carried out. A large-scale plan involving 32 trials was also drawn up to identify possible interactions, which, if undetected, could lead to false conclusions. Evaluation of the results gave the following order of importance:

C. Diameter of discs
M. Type of grinding media
B. Size of vessel
F. Pigment concentration
H. Pre-mixing stage
G. Pigment/binder ratio

This permits the interesting conclusion to be drawn that of the six significant factors, the three most important are related to the dispersing machine whereas the remaining less important factors concern the mill base and its preparation.

In further series of tests, the significance of the interactions and then the optimum levels of the significant factors were determined. However, we will dispense with a detailed discussion at this point because our purpose here is solely to point out the possibilities of statistical experimental designs for the optimisation of mill base compositions. The interested reader will find a thorough description of these experimental plans with discussion of the results and conclusions in the paper by *Carr* and *Kelly* [176].

Important conclusions from the series of tests that followed were:

1. The effect of disc speed passed through a maximum lying at 3000 rpm. At this speed, the maximum tinting strength was attained after 20 minutes grinding time. Longer grinding did not increase the tinting strength.
2. Presumably the most important conclusion: degree and rank order of the significant factors were specific to the medium and pigment, as could be found with a series of tests using other pigments (C.I. Pigment Orange 5; N° 12075 and Pigment Red 48:4; N° 15865:4) in the same medium, and then in other media.
3. Sand was far more suitable than glass beads. This turned out to be the case with all tested pigments.

In addition, it could also be shown that all conclusions were only valid for the ranges of variance investigated. The general validity thus remains to be tested.

The fact that the effect of factors governing the dispersion process is not only specific to the medium, but also to the pigment has been confirmed by practical experience. Efforts made to correlate the dispersing effects of several types of milling devices for given media failed, as this correlation also proved to be pigment-specific.

Further application of statistical experimental plans to the optimisation of mill base compositions are discussed in articles by *Gall* and *Siebke* [177, 178].

6.4.5 Optimisation of the cost of the dispersion process by controlling energy input

Before concluding this chapter, a word must be said regarding the possible optimisation of costs by reducing the power consumption. In ball mills of the stirrer-type, the energy input required for the dispersion process constitutes only a small fraction of the total energy consumed in the stirring process, as the greater part is lost to the cooling water as frictional heat. However, this heat energy carried off in the cooling water can occasionally be recovered and used again.

Winkler and *Dulog* investigated the energetic/economic aspect of pigment dispersion in ball mills of the stirrer-type by using grinding devices especially equipped with measuring instruments [179]. They followed the progress of the dispersion by measuring the development of various parameters and the stirring performance. The curves obtained for the parameters (tinting strength, gloss, fineness of grind and viscosity) as functions of the mechanical stirring performance, were similar to the corresponding curves obtained as functions of the milling time. The optimum values for stirring performance with regard to gloss, tinting strength and fineness of grindwere of a similar order of magnitude. Viscosity, on the other hand, required higher performance.

In view of the observation that it makes no difference to the attained degree of dispersion whether the stirring power is set by adjusting the torque or the number of revolutions, it would seem advisable to carry out the dispersion process using as moderate a flow of cooling water as possible, in order to sink the energy costs. Thus, the torque level in the mechanical stirring power is lower and, on the other hand, the water consumption is reduced. If there are no coloristic reasons speaking against a higher abrasion of the grinding media, the mill base should be taken to the limits of its thermal endurance during dispersion.

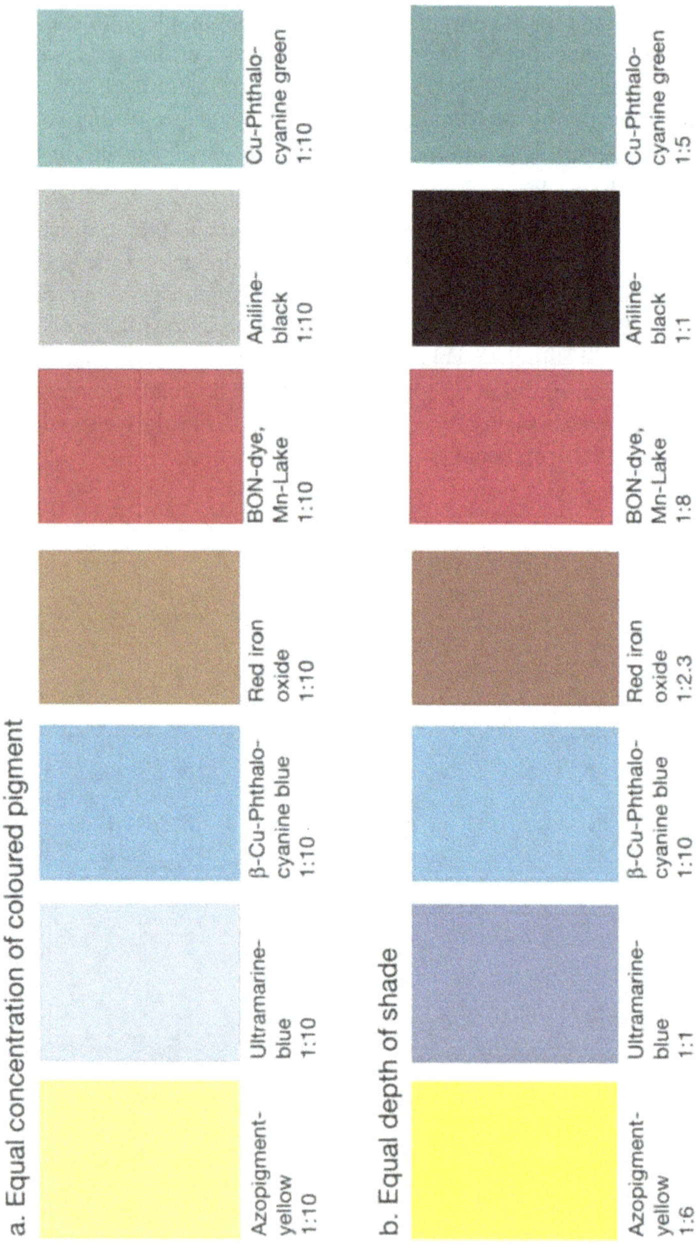

Figure 2.16: Different inorganig and organic coloured pigments at equal concentration (a) and equal depth of shade (b)

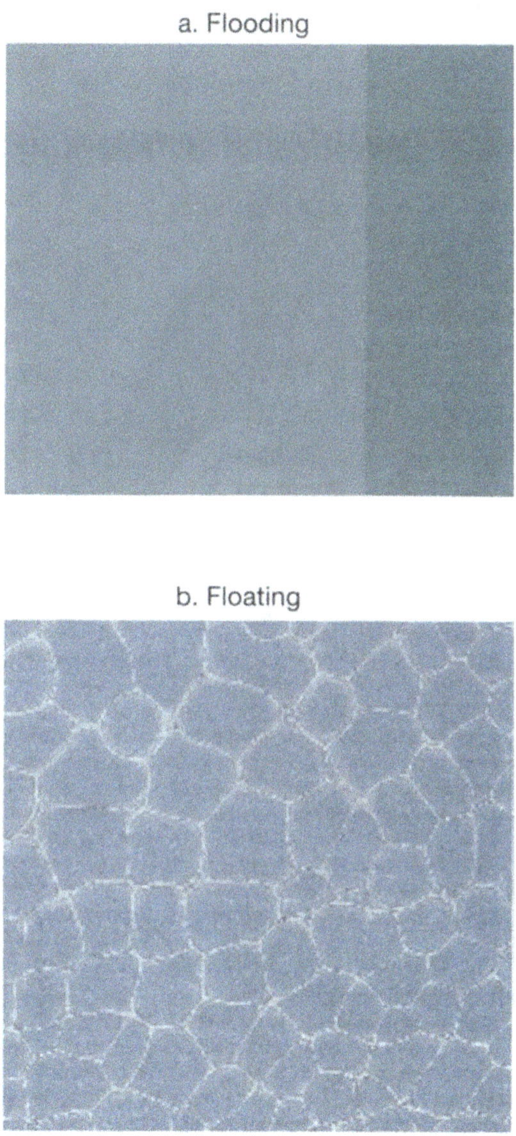

Figure 4.16: Reproductions of coating surfaces; a: flooding (rub-out effect), b: floating (Bénard cells)

Appendix

List of physical symbols and constants used

Symbol	Meaning
A	surface
$\mathrm{A_{ad}}$	work of adhesion
$\mathrm{A_H}$	Hamaker constant
$\mathrm{A_{co}}$	work of cohesion
B	mobility of particles
B	binder concentration
c	concentration
c	speed of sound in liquid
$\mathrm{c_V}$	number of particles per volume
d	lattice constant of a crystal
d	diameter
d_E	equivalent diameter
D	velocity gradient
D	dispersibility
D^*	diameter of the optimum particle size
DH	dispersion hardness
DL	lightness flop
e	Euler number **or** electrical elementary charge
E	field strength **or** energy
E_A	attractive energy
E_R	repulsive energy
F	surface of a sphere **or** reflection function
F	force **or** free energy **or** tinting strength
F_a	absorption factor
$\mathrm{F_{Ru}}$	correction factor for surface rugosity
F_s	scattering factor
F/C	Guggenheim factor
g	gravitational acceleration
G	shear modulus **or** free enthalpy
G_α	inertia term
h	height **or** Planck constant
h	layer thickness **or** concentration in DIN 55 988
H	interaction energy

H	enthalpy
i_1	vertically polarised component of the scattered radiation
i_2	horizontally polarised component of the scattered radiation
IS	increase in tinting strength
I_0	intensity of the primary radiation
I_i	intensity of the incident radiation
I_r	Intensity of the reflected radiation
I_s	intensity of the scattered radiation
J	intensity of the light reflected back
k	Boltzmann constant
k or K	constant
K	absorption coefficient **or** coarseness
K*	conductivity of suspension
K_{spr}	spreading coefficient
l	length
m	weight amount
n	integer for the order of reflection in a crystal lattice **or** ion concentration
n	refractive index
N	number of particles **or** molecules per unit volume **or** unit mass
N	degree of efficiency
N_A	Avogadro number
p	pressure
p_i	induced dipole moment
P	probability **or** pigment concentration **or** pressure wave
P_s	saturation vapour pressure
q	Mie scattering parameter
q	number of atoms **or** molecules per millilitre
Q_a	absorption cross-section
Q_s	scattering cross-section
r or R	radius
r_0	coefficient of external reflection
r_2	coefficient of internal reflection
R	universal gas constant **or** reflectance
R_s	degree of reflection of light polarised vertically to the incident plane
R_p	degree of reflection of light polarised parallel to the incident plane
R_1	degree of reflectance at oblique light incidence
R_2	degree of reflectance at vertical light incidence
R_∞	reflectance of an infinitely thick layer
R_λ	reflection factor
S	entropy **or** surface **or** scattering coefficient
S_λ	radiation intensity in the CIE system

t	time
T	transparency number
T	thermodynamic temperature
v	sedimentation velocity
V	volume
x	path length, distance
z	valency of the ions
α	molecular polarisability
α	angle of incidence
β	refracting angle **or** London constant
γ	interface tension
δ	solubility parameter
ΔE	molar activation energy
ΔE_{ab}*	hue difference in CIELab system
ΔF	difference in depth of shade
ΔY_{∞}	difference in lightness
Δρ	density difference
ε	extinction coefficient
ε	dielectric constant
ε_{el}	elastic deformation
ζ	zeta potential
η	viscosity coefficient
η_{red}	reduced viscosity
η_{rel}	relative viscosity
η_s	specific viscosity
η_{∞}	final viscosity
ϑ	Bragg's glancing angle
θ	surface of adsorbent covered with absorbate
κ	reciprocal layer thickness
λ	wave length
λ_M	wave length in the medium
μ	electronic mobility
μ_D	electrophoretic mobility
ρ	density
σ	surface tension **or** bond energy
τ	shear stress
τ_0	yield value
φ	viewing angle
ψ_0	electrostatic potential on the particle surface
ω	angle of contact
ω	angular velocity

List of relevant norms

DIN 1342-1, *Viscosity – Part 1: Rheological concepts*

DIN 13900-2, *Pigments and extenders – Methods of dispersion and assessment of dispersibility in plastics – Part 2: Determination of colouristic properties and ease of dispersion in plasticized polyvinylchloride by two-roll milling; German version EN 13900-2:2003*

DIN 53019-1, *Viscometry – Measurement of viscosities and flow curves by means of rotational viscometers – Part 1: Principles and measuring geometry*

DIN 53164, *Testing of pigments and extenders; determination of relative scattering power of white pigments – Black-ground method*

DIN 53222, *Determination of viscosity by falling-rod viscometer* ((withdrawn 3023-12))

DIN 53235-1, *Testing of pigments – Tests on specimens having standard depth of shade – Part 1: Standard depths of shade*

DIN 53235-2, *Testing of pigments – Tests on specimens having standard depth of shade – Part 2: Adjustment of specimens to standard depth of shade*

DIN 53238 (all parts), *Pigments and extenders – Method of assessment of ease of dispersion*

DIN 53238-22, *Testing of pigments; test of ease of dispersion, determination of the development of tinting strenght and hardness of dispersion* (replaced by DIN EN ISO 8780-1)

DIN 55601, *Testing of pigments – Determination of a hiding power value of dried paints using a wedge-shaped layer*

DIN 55982, *Testing of pigments – Determination of lightening power of white pigments – Paste method*

DIN 55987, *Testing of pigments – Determination of hiding power value of pigmented media – Colorimetric method*

DIN 55988, *Determination of indices for the transparency of pigmented and unpigmented systems – Colorimetric method*

DIN 66126-1, *Determination of specific surface area of disperse solids by the permeability technique – Fundamentals*

DIN 66126-2, *Determination of specific surface area of disperse solids by the permeability technique – Blaine method and apparatus*

DIN 66143, *Graphical representation of particle size distributions – Power-function grid*

DIN 66144, *Graphical representation of particle size distributions – Log-normal grid*

DIN EN ISO 787-5, *General methods of test for pigments and extenders – Part 5: Determination of oil absorption value (ISO 787-5:1980); German version EN ISO 787-5:1995*

DIN EN ISO 787-7, *General methods of test for pigments and extenders – Part 7: Determination of residue on sieve – Water method – Manual procedure (ISO 787-7:2009); German version EN ISO 787-7:2009*

DIN EN ISO 787-16, *General methods of test for pigments and extenders – Part 16: Determination of relative tinting strength (or equivalent colouring value) and colour on reduction of coloured pigments – Visual comparison method (ISO 787-16:1986); German version EN ISO 787-16:1995*

DIN EN ISO 787-18, *General methods of test for pigments and extenders – Part 18: Determination of residue on sieve – Mechanical flushing procedure (ISO 787-18:1983); German version EN ISO 787-18:1995*

DIN EN ISO 787-24, *General methods of test for pigments and extenders – Part 24: Determination of relative tinting strength of coloured pigments and relative scattering power of white pigments – Photometric methods (ISO 787-24:1985); German version EN ISO 787-24:1995*

DIN EN ISO 1524, *Paints, varnishes and printing inks – Determination of fineness of grind (ISO 1524:2013); German version EN ISO 1524:2013*

DIN EN ISO 2813, *Paints and varnishes – Determination of gloss value at 20°, 60° and 85° (ISO 2813:2014); German version EN ISO 2813:2014*

DIN EN ISO 3219, *Plastics – Polymers/resins in the liquid state or as emulsions or dispersions – Determination of viscosity using a rotational viscometer with defined shear rate (ISO 3219:1993); German version EN ISO 3219:1994*

DIN EN ISO 4618, *Paints and varnishes – Terms and definitions (ISO 4618:2014); Trilingual version EN ISO 4618:2014*

DIN EN ISO 6504-1, *Paints and varnishes – Determination of hiding power – Part 1: Kubelka-Munk method for white and light-coloured paints (ISO 6504-1:1983); German version EN ISO 6504-1:2006*

DIN EN ISO 8780-1, *Pigments and extenders – Methods of dispersion for assessment of dispersion characteristics – Part 1: Introduction (ISO 8780-1:1990); German version EN ISO 8780-1:1995*

DIN EN ISO 8780-3, *Pigments and extenders – Methods of dispersion for assessment of dispersion characteristics – Part 3: Dispersion using a high-speed impeller mill (ISO 8780-3:1990); German version EN ISO 8780-3:1995*

DIN EN ISO 8781-1, *Pigments and extenders – Methods of assessment of dispersion characteristics – Part 1: Assessment from the change in tinting strength of coloured pigments (ISO 8781-1:1990); German version EN ISO 8781-1:1995*

DIN EN ISO 8781-2, *Pigments and extenders – Methods of assessment of dispersion characteristics – Part 2: Assessment from the change in fineness of grind (ISO 8781-2:1990); German version EN ISO 8781-2:1995*

DIN EN ISO 8781-3, *Pigments and extenders – Methods of assessment of dispersion characteristics – Part 3: Assessment from the change in gloss (ISO 8781-3:1990); German version EN ISO 8781-3:1995*

DIN EN ISO 11664-4, *Colorimetry – Part 4: CIE 1976 L*a*b*Colour space (ISO 11664-4:2008); German version EN ISO 11664-4:2011*

DIN EN ISO 13803, *Paints and varnishes – Determination of haze on paint films at 20° (ISO 13803:2014); German version EN ISO 13803:2014*

DIN ISO 9276-1, *Representation of results of particle size analysis – Part 1: Graphical representation (ISO 9276-1:1998)*

DIN ISO 9277, *Determination of the specific surface area of solids by gas adsorption – BET method (ISO 9277:2010)*

DIN ISO 18451-1, *Pigments, dyestuffs and extenders – Terminology – Part 1: General terms (ISO/DIS 18451-1:2014)*

DIN ISO 18451-2, *Pigments, dyestuffs and extenders – Terminology – Part 2: Classification of colouring materials according to coloristic and chemical aspects (ISO/DIS 18451-2:2014)*

ISO 2846-1, *Graphic technology – Colour and transparency of printing ink sets for four-colour printing – Part 1: Sheet-fed and heat-set web offset lithographic printing*

ISO 12644, *Graphic technology – Determination of rheological properties of paste inks and vehicles by the falling rod viscometer*

ISO 13320, *Particle size analysis – Laser diffraction methods*

ASTM D 523, *Standard Test Method for Specular Gloss*

ASTM D 869, *Standard Test Method for Evaluating Degree of Settling of Paint*

ASTM D 1210, *Standard Test Method for Fineness of Dispersion of Pigment-Vehicle Systems by Hegman-Type Gage*

ASTM D 1483, *Standard Test Method for Oil Absorption of Pigments by Gardner-Coleman Method*

ASTM D 2805, *Standard Test Method for Hiding Power of Paints by Reflectometry*

References

[1] *B. Honigmann*, "Aufbau und Abmessungen der Teilchen gut kristallisierter organischer Pigmente", Farbe & Lack, Vol. 82, N°9 (1976), p. 815-820

[2] *W. Herbst* and *K. Hunger,* "Industrial Organic Pigments" VHC Verlagsgesellschaft, Weinheim, 1993, p. 8 and p. 451-455

[3] *J. Schröder*, "Morphologie organischer Pigmente unter besonderer Berücksichtigung der Kupferphthalocyanine", Farbe & Lack, Vol. 90, N° 1 (1984), p. 6-18

[4] *P. W. Atkins*, "Physical Chemistry", Oxford University Press, 5th Edition, Oxford, 1997, S. 727-732

[5] *P. Hauser* and *B. Honigmann,* "Eine Scheibenzentrifuge mit laser-optischer Einrichtung zur Teilchengrößenanalyse von Pigmentdispersionen", Farbe & Lack, Vol. 83, N° 10 (1978), p. 886-890

[6] *W. Carr*, "Assessment of Pigment Dispersion", Progress in Organic Coatings, N°4 (1976), p. 161-188

[7] *R. R. Irani* and *F.C. Callis,* "Particle Size: Measurement, Interpretation and Application" John Wiley, New York, 1963

[8] *R.D. Cadle*, "Particle Size. Theory and Industrial Application" Reinhold Publishers, New York, 1965

[9] *H. Sonntag*, "Lehrbuch der Kolloidwissenschaft", VEB Deutscher Verlag der Wissenschaften, Berlin, 1977, p. 162

[10] *D. Patterson*, "Pigments: An Introduction to their Physical Chemistry", Elsevier Publishing Co. Ltd., Barking, Essex and New York, 1967

[11] S. *Brunauer, P.H. Emmett* and *E. Teller,* "Adsorption of Gases in Multimolecular Layers", Journal of the American Chemical Society, Vol. 60, N° 2 (1938), p. 309-319

[12] *K. Apel*, "Beziehungen zwischen spezifischer Oberfläche und anwendungstechnischen Eigenschaften bei organischen Pigmenten", IXth FATIPEC Congr., Bruxelles 1968, Congress Book, p. 28-33

[13] *Leslie S. Ettre,* "Measurement and Characterisation by Gas Absorption", in T.C. Patton, "Pigment Handbook", John Wiley, New York, 1988, Vol. III-B, p. 139-144

[14] *A.C. Zettlemoyer*, "The Pigment – Vehicle Interface", Official Digest, December 1957, p. 1238-1271

[15] *O. J. Schmitz, P. J. Sell* and *K. Hamann,* "Kennzeichnung der Hydrophilität von Pigmentoberflächen durch Benetzungswärmen", Farbe & Lack, Vol. 79, N° 11 (1973), p. 1049-1058

[16] *R. Sappock*. "Recent progress in the physical-chemistry of organic pigments with special reference to phthalocyanines", J. Oil Col. Chem. Assoc., Vol. 61, N° 8 (1978), p. 299-308

[17] *H. D. Dörfler*, "Grenzflächen- und Kolloidchemie", VCH Verlagsgesellschaft, Weinheim, 1994, p. 18-34

[18] *W. A. Zisman,* "Adhesion and Cohesion", P. Weiss Verlag Elsevier, New York, 1962, p. 176

[19] *U. Zorll*, "Bedeutung und Problematik der kritischen Oberflächenspannung", Adhäsion, N° 9 (1974), p. 262-270

[20] *P. J. Sell* and *A.W. Neumann,* "Die Oberflächenspannung fester Körper", Angewandte Chemie, Vol. 78, N°6 (1966), p. 321-331

[21] *G. E. Hellwig* and *A.W. Neumann,* "Randwinkelmessungen zur Bestimmung von Benetzungsenergien im Hinblick auf das Verhalten von Pigmenten", Farbe & Lack, Vol. 73, N°. 9 (1967), p. 823-829

[22] *P. J. Sell* and *D. Renzow,* "Nomographische Ermittlung von Randwinkeln und Festkörperoberflächenspannungen an Kupferphthalocyanin", Farbe & Lack, Vol. 83, N°4 (1977), p. 265-269

[23] *P. J. Sell* and *A. W. Neumann,* "Ein Verfahren zur Bestimmung der Oberflächenspannungen und Grenzflächenspannungen von Festkörpern", Zeitschrift für physikalische Chemie, Neue Folge, Vol. 41 (1964), p. 191-196

[24] anon., "Kronos Guide", Editor: Kronos Titanium Companies, Leverkusen, 1968, p. 38

[25] *H. Ferch,* "Pigmentruße für leitfähige Beschichtungen", XIXth FATIPEC Congr., Aachen 1988, Congress B. II, p. 141-176

[26] *K. Merkle* and *H. Schäfer,* "Surface Treatment of Pigments", in T. C. Patton "Pigment Handbook", John Wiley, New York, 1988, Vol. III-B, p. 157-167

[27] *J. Schroeder,* "Oberflächenbelegung von Pigmenten", Farbe & Lack, Vol. 93, N° 9 (1987), p. 715-720.

[28] *P. Bugnon,* "Surface treatment of pigments. Treatment with inorganic materials" Progress in Organic Coatings, Vol. 29, N° 1 (1996), p. 40-43

[29] U. Zorll, "New Aspects of Gloss of Paint Film and its Measurement", Progress in Organic Coatings, 1 (1972), p. 113-155

[30] *B. Dicke and F. Fensterseifer,* "Glanzmessung in Theorie und Praxis", a Technical Information from Byk Gardner GmbH, Geretsried (Germany), 1989

[31] anon., "Application" 1 and 2, Technical Informations from Byk Gardner GmbH, Geretsried (Germany), 1990

[32] *F. Fensterseifer B. Dicke* and *W. Stahl,* "Einfach glänzend", Journal für Oberflächentechnik, N° 6 (1995), p. 68

[33] *L. Dulog* and *U. Wustmann,* "Glanzschleier und die Pigment/Bindemittelwechselwirkung", Farbe & Lack, Vol. 102, N° 2 (1996), p. 28-35

[34] G. Mie, "Beiträge zur Optik trüber Medien, speziell kolloidaler Metallösungen", Annalen der Physik, Vol. 25 (1908), p. 377-445

[35] *H. C. van de Hulst,* "Light Scattering by Small Particles", John Wiley & Sons, New York, 1957

[36] *Max Born,* "Principles of Optics", Pergamon Press, Oxford and London, 4th edition, 1970, p. 662

[37] *Bergmann and Schaefer,* "Lehrbuch der Experimentalphysik", 8th Edition, Vol. 3 Optik, Walter de Gruyter, Berlin, 1987, p. 271

[38] *H. H. Weber* "Über das optische Verhalten von kugeligen, isotropen Teilchen in verschiedenen Medien", Kolloid-Zeitschrift, Vol. 174, N° 1 (1961), p. 66-72

[39] *R. Gans,* "Über die Form ultramikroskopischer Goldteilchen", Annalen der Physik", Vol. 37, 1912, p. 881-900

[40] *P. Kubelka* and *F. Munk,* "Ein Beitrag zur Optik der Farbanstriche", Zeitschrift für technische Physik", N° 11a (1931), p. 593-601

[41] H. Völz, "Industrial Colour Testing", VCH Verlagsgesellschaft, Weinheim, 1995

[42] L. Gall, "The Colour Science of Pigments", a monograph from Badische Anilin- & Soda-Fabrik AG, Ludwigshafen, 1972
[43] *H. G. Völz,* "Bestimmung des Aufhellvermögens in Abhängigkeit der PVK des Weißpigments", Farbe & Lack, Vol. 88, N° 5 (1982), p.356-360
[44] *M. A. Maikowski* "Lichtabsorption in pigmentierten Schichten", Report of the Bunsengesellschaft, Vol. 71, N° 3 (1967), p. 313-326
[45] *L. Gall* and *G. Riedel,* "Die Bedeutung von Standardfarbtiefen und deren farbmetrische Bestimmung", Farbe & Lack, Vol. 71, N° 11 (1965), p. 894-901
[46] *W. Kaufmann,* "Erfahrungen mit feinst vermahlenen Pigmenten bei der Verarbeitung zu Anstrichmitteln", Farbe & Lack, Vol. 62, N° 9 (1956), p. 428-431
[47] *W. Carr,* "Effect of Pigment Dispersion on the Appearance and Properties of Paint Films", J. Oil. Chem. Assoc., Vol. 54, N° 12 (1971), p. 1093-1127
[48] *P. Hauser, B. Honigmann* and *M. Herrmann,* "Abhängigkeiten koloristischer und rheologischer Eigenschaften pigmentierter Leinölfirnisse von Teilchengröße und Verteilung der Pigmente", Farbe & Lack 76 Vol. 76, N°6 (1970), p. 545-550 and Vol. 77, N° 11 (1971), p. 1097-1106
[49] *G. Kämpf,* "Korngrößenverteilung und optische Eigenschaften von Pigmenten", Farbe & Lack, Vol. 71, N° 5 (1965), p. 353
[50] *A. Brockes,* "Der Zusammenhang von Farbstärke und Teilchengröße von Buntpigmenten nach der Mie-Theorie", Optik, Vol. 21 (1964), p. 550-566
[51] *M.A. Maikowski,* "Der Einfluß der Korngrößenverteilung auf koloristische Eigenschaften und Lichtechtheit organischer Pigmente", Farbe & Lack, Vol. 7, N° 7 (1971), p. 640-647
[52] *W. Herbst* and *O. Hafner,* "Untersuchungen über die Pigmentierung moderner Lacksysteme", Farbe & Lack, Vol. 82, N° 5 (1976), p. 394
[53] *L. Dulog* and *H. Schweiger,* "Pigmente in elektronenstrahlhärtenden Systemen", Farbe & Lack, Vol. 103, N° 8 (1997), p. 30-44
[54] *M. Pirani* and *J. Fischer,* "Graphische Darstellung in Wissenschaft und Technik", Sammlung Göschen, Vol. 728/728a, Walter de Gruyter, Berlin, 1957, p. 20
[55] *W. Wittenberger,* "Rechnen in der Chemie", Springer Verlag, Wien, 1960, p. 80
[56] *N. Casson,* "Rheology of Disperse Systems", C. C. Mill Editor, Pergamon Press, New York 1959, p. 84-104
[57] *M. Reiner,* "Rheologie in elementarer Darstellung", Carl Hanser Verlag, München, 1968, p. 264-266
[58] *D. Hadjistamov,* "Indirekte Methoden zur Ermittlung der Fließgrenze bzw. Fließgrenzbereiche", Farbe & Lack, Vol. 91, N° 11 (1985), p. 1023-1029
[59] *H. Schmidbauer,* "Praxisorientierte Rheologiemessungen von modernen Lacksystemen", XXth FATIPEC Congr., Nizza 1990, Congress Book, p. 429-433
[60] *W. Moore,* "Physical Chemistry", Longmans Green & Co. Ltd., London 1968, p. 776-777
[61] *H. J. Adler,* "Rheological investigations into water-thinnable paint binders" Rheology, Vol. 92, June (1992), p. 96-101
[62] *P. E. Pierce,* "Rheology of Coatings", Journal of Coatings Technology, Vol. 41, N° 533 (1969), p. 383-395
[63] *A. C. Zettlemoyer* and *G. W. Lower,* "The Rheology of Printing Inks. III. Studies of Simple Dispersion", Journal of Colloid Science, Nr. 10 (1955), p. 29-45

[64] *F. R. Eirich,* "Rheology. Theory and Applications", Academic Press, New York and London 1960, Vol. 3, Chapt. 5 The Rheology of Printing Inks, p. 145

[65] *A. C. Zettlemoyer,* "Rheology of Printing Inks", American Ink Maker, Nr. 2 (1953), p. 31-33 and 63-65

[66] *H. Pahlke,* "Untersuchungen über das Fließverhalten von Druckfarben", Farbe & Lack, Vol. 75, N° 3 (1969), p. 236-243

[67] anon., "Leitfaden für den Einsatz rheologischer Additive", Technical Information from *Kronos Titan GmbH,* 1980

[68] *K. Merkle, W. Herbst* and *G. Eulitz,* "Neue organische Volltonpigmente für deckende bleifreie Pigmentierungen", Farbe & Lack, Vol. 82, N° 9 (1976), p. 801-804

[69] *F. Molenaar* and *M. Hoeflaak,* "Viscoelastic Properties of Coatings", XXth FATI-PEC Congr., Nizza 1990, Congress Book p. 159-167

[70] *U. Zorll,* "Rheologische Begriffe zur Charakterisierung des Fließverhaltens von Lackmaterialien", Farbe & Lack, Vol. 86, N° 4 (1980), p. 301-307

[71] *T. C. Patton,* "Fundamentals of Paint Rheology", Journal of Paint Technology, Vol. 40, N° 522, (1968), p. 301-307

[72] *W. M. Kulicke,* "Fließverhalten von Stoffen und Stoffgemischen", Hüthig & Wepf, Basel and Heidelberg, 1986, p. 267

[73] *T. C. Patton,* "Paint Flow and Pigment Dispersion", John Wiley & Sons Inc., New York 1979, p. 14-79

[74] *T. Mezger,* "Ein kleiner Rheologiekurs", a company booklet from Physika Messtechnik GmbH, 1991

[75] *G. Schramm,* "Introduction to Practical Viscometry", a company booklet from Gebr. Haake GmbH, Stuttgart, 1991

[76] *H. Schäfer,* "The Influence of the dispersion process on the colouristic and fastness properties of molybdate red pigments", reprint of Hoechst AG in English, originally published in German in Farbe & Lack, Vol. 77, N°. 11 (1971), p. 1081-1090

[77] *H. Rumpf,* "Haftung und Festigkeit von Agglomeraten – Vergleich zwischen Modellrechnung und Experiment", Die pharmazeutische Industrie, Vol. 34, N° 4 (1972), p. 270-281

[78] *T. C. Patton,* "Paint Flow and Pigment Dispersion", John Wiley & Sons, Inc. New York, 1979, p. 376-387

[79] *H. Rumpf,* "Die Einzelkornzerkleinerung als Grundlage einer technischen Zerkleinerungswissenschaft", Chemie Ingenieur Technik, Vol. 37, N° 3 (1965), p. 187-202

[80] *G. Parfitt,* "Dispersion of Powders in Liquids", Applied Science Publishers, London and New Jersey, 3rd Edition, 1981, p. 9

[81] *G. Parfitt,* "Dispergierung von Pigmenten in Theorie und Praxis", Deutsche Farben Zeitschrift, Vol. 32, N° 9 (1978), p. 322-331

[82] *E. Washburn,* "The Dynamics of Capillary Flow", Physical Review, Vol. 18, Ser. 2 (1921), p. 273-283

[83] *D. H. Napper,* "Polymeric Stabilisation of Colloidal Dispersions", Academic Press, London, New York and Sydney, 1983. p. 144-159

[84] *G. Joppien,* "Strukturuntersuchungen an Adsorptionsschichten makromolekularer Stoffe", Die makromolekulare Chemie, Vol. 175 (1974), p. 1931-1954 and Vol. 176 (1975), p. 1129-1149

[85] *G. Joppien* and *K. Hamann,* "The Structure of Layers of Adsorbed Polymers at Pigment/Solution-Interfaces and their Influence on the Dispersion Stability of Pigments in Paints", J. Oil Col. Chem. Assoc., Vol. 60, N° 10 (1977), p. 412-423

[86] *K. Hamann,* "Über die Wechselwirkung zwischen Pigmenten und Bindemitteln", VIth FATIPEC Congr., Wiesbaden 1962, Congress Book, p. 76-85

[87] *K. Řeháček,* "Untersuchung der Wechselwirkung zwischen Pigment und Bindemittel in konzentrierten Bindemittellösungen", Farbe & Lack, Vol. 76, N° 7 (1970), p. 656-665

[88] *E. C. Rothstein,* "Role of the Absorbed Resin Layer in the Dispersion of Pigmented Alkyd Systems", Official Digest, Vol. 36, (1964), p. 1448

[89] *W. D. Schaeffer,* "A Supplementary Analysis of Adsorbed Alkyd Films" Official Digest, Vol. 37 (1965), p. 1305

[90] *T. Sato,* "Adsorption of Polyamides and the Stability of Dispersion", Journal of Applied Polymer Science, Vol. 15, N° 5 (1971), p. 1053-1067

[91] *P. Sørensen* "Application of the Acid/Base Concept Describing the Interaction between Pigments, Binders and Solvents", Journal of Paint Technology, Vol. 47, N° 3 (1975), p. 31-39

[92] *H. Dittrich, E. Hergesell and K. M. Oesterle,* "Zur direkten Ermittlung günstiger Dispergieransätze", *E. Hergesell* and XIIth FATIPEC Congr., Garmisch-Partenkirchen 1974, Congress Book, p. 255-262

[93] *W. Carr,* "The Theory and Practice of Pigment Dispersion", J. Oil Col. Chem. Assoc., Vol. 61, N° 11 (1978), p. 397-410

[94] *J. Hildebrand,* "Solubility", Journal of the American Chemical Society, Vol. 38 (1916), p. 1452-1473

[95] *H. Burrell,* "Solubility Parameters for Film Formers", Official Digest, Vol. 27, N° 369 (1955), p. 726

[96] *H. Burrell,* "The Challenge of the Solubility Parameter Concept", Journal of Paint Technology, Vol. 40, N° 5 (1968), p. 197-207

[97] *P. Sørensen,* "Der Begriff des Löslichkeitparameters in Pigment-Harz-Lösungsmittelsystemen", Deutsche Farben Zeitschrift, Vol. 24, N° 10 (1970), p. 473-479

[98] *R. L. Eissler, R. Zgol* and, *J. A. Stolp,* "Solubility Parameters Applied to Sedimentation Volumes of Pigments", Journal of Paint Technology, Vol. 42, N° 9 (1970 p. 483-489

[99] *L. Brandts-Buys* and *P. Huyskens,* "Limitations dans l'applicabilité du concept des paramètres de solubilité", XVIth FATIPEC Congr., Liège 1982, Congress Book Vol. II, p. 191-204

[100] *V. T. Crowl,* "Floculation, flotation and flooding in phthalocyanine/titanium dioxide pigmented paints", J. Oil Col. Chem. Assoc., Vol. 50, N° 11 (1967), p. 1023-1059

[101] *M. J. Vold,* "The Effect of Adsorption on the van der Waals Interaction of Spherical Colloidal Particles", J. Colloid Science, Vol. 16, N° 1 (1961), p. 1-12

[102] *E. J. Clayfield* and *E. C. Lumb,* "A Theoretical Approach to Polymeric Dispersant Action", J. Colloid Interface Science, Vol. 22, N° 3 (1966) p. 269-293

[103] *R. I. Feigin* and *D. H. Napper,* "Depletion Stabilisation and Depletion Flocculation", J. Colloid Interface Science, Vol. 75, N° 2 (1980), p. 525-541

[104] *B. V. Deryaguin, L. D. Landau,* Acta Phisicochim and *L. D. Landau,* URSS, Vol. 14 (1941), p. 633

[105] *E. J. W. Verwey* and *J. G. Overbeek,* "Theory of the Stability of Liophobic Colloids", Elsevier, Amsterdam, 1948

[106] *F. Daniel,* "Flocculation and Related Pigment Interactions in Paints", Journal of Paint Technology, Vol. 38, N° 497, N°6 (1966), p. 309-316.
[107] *P. Kresse,* "Eine chromatographische Methode zur Bestimmung der Flockulation von Pigmenten", Farbe & Lack, Vol. 76, N° 4 (1970), p. 351-358
[108] *S. Keifer,* "Streuvermögen weißer Anstrichfarben als Maß für den Flockungsgrad", Farbe & Lack, Vol. 79, N° 12 (1973), p. 1161-1166
[109] *J. G. Balfour* and *M. J. Hird,* "Flocculation: Its Measurement and Effect on Opacity in Systems Containing Titanium Dioxide Pigments", J. Oil Col. Chem. Assoc., Vol. 58, N° 9 (1975), p. 331-344
[110] *M. v. Smoluchowski,* "Versuch einer mathematischen Theorie der Koagulationskinetik kolloider Lösungen", Zeitschrift für physikalische Chemie, Vol. 92, N° 2 (1918), p. 129-168
[111] *F. Daniel,* "Flocculation and Related Pigment Interactions in Paints", VIIth FATIPEC Congr., Vichy 1964, Congress Book, p. 280-288
[112] *O. J. Schmitz, L. Dulog* and *R. Fischer,* "Untersuchungen zur Stabilität von Pigmenten in umweltfreundlichen, wäßrigen Lacksystemen", XXIst FATIPEC Congr., Amsterdam 1992, Congress B., Vol. 1, p. 89
[113] *W. Herbst* and *K. Hunger,* "Industrial Organic Pigments", VCH Verlagsgesellschaft, Weinheim, 1993, p. 429-430
[114] *J. Ruf* "Organischer Metallschutz", Vincentz Verlag, Hannover, 1993, p. 415-418
[115] *J. Schroeder,* "Säure-Base-Eigenschaften von Pigmenten", Farbe & Lack, Vol. 96, N° 2 (1990), p. 91-97
[116] *P. Kresse,* "Einfluß von Teilchenform und Teilchengröße anorganischer Pigmente auf Ausschwimmerscheinungen", Farbe & Lack, Vol. 72, N° 2 (1966), p. 111-118
[117] *T. C. Patton,* "Paint Flow and Pigment Dispersion", John Wiley & Sons, Inc. New York, 1979, p. 508-510
[118] *U. Kaluza,* "Über das Rub-out-Verhalten pigmentierter Systeme", Deutsche Farben Zeitschrift, Vol. 27, N° 9 (1973), p. 427-439
[119] *A. Topham,* „Dispersing Agents for Pigments in Organic Liquids", Progress in Organic Coatings, N° 5 (1977), p. 237-243
[120] *E. Karsten,* "Lackrohstofftabellen", Vincentz Verlag, Hannover, 10th Edition, 2000, p. 65
[121] *P. Kresse,* "Farbänderungen an Lackfilmen durch Ausschwimmen, Flockulation, Sedimentation und ihre Verhinderung", Deutsche Farben Zeitschrift, Vol. 24, N° 11 (1970), p. 521-526
[122] *J. Winkler,* "Über das flockungsstabilisierende Verhalten feinteiliger Blanc fixe", Farbe & Lack, Vol. 97, N° 10 (1991), p. 859-866
[123] *U. Kaluza,* "Flocculation of Pigments in Paints – Effects and Causes", Progress in Organic Coatings, N° 10 (1982), p. 289-330
[124] *F. Haselmeyer,* "Theorie und Praxis des Ausschwimmens von Pigmentkombinationen in Lacken", Farbe & Lack, Vol. 74, N° 7 (1968), p. 662-668
[125] *W. Krauß,* "Die Anwendung der Silikone in der Lackindustrie", Farbe & Lack, Vol. 64, N° 6 (1958), p. 309-315
[126] *R. Haug,* "Über Seeding-Effekte und ihre Vermeidung", Meeting of the Gesellschaft Deutscher Lackforschung e. V., Stuttgart, April 1967
[127] *H. Sander,* "Über den Pigment-, Lösungsmittel- und Bindemittelschock", Farbe & Lack, Vol. 77, N° 9 (1971), p. 891-892

[128] anon., “Schockeffekte, ihre Ursache und Vermeidung”, Technical Information from Farbenfabriken Bayer A.G., 1992

[129] *W. Liehr,* “Einige Untersuchungen zum Problem des chemischen Seeding”, Farbe & Lack, Vol. 74, N° 1(1968), p. 31-37

[130] *W. Garmsen,* “Kornfeinheitsmessungen an Lackfarben”, Farbe & Lack, Vol. 59, N° 3 (1953), p. 92-97

[131] *A. Becker et al.,* “Optoelektronisches Gerät zur Beurteilung von Pigmentdispersionen”, Farbe & Lack, Vol. 97 N° 11 (1991), p. 972-975

[132] *O. Schmitz,* “Kennzahlen für die Dispergierbarkeit von Pigmenten”, XIIth FATIPEC Congr, Garmisch-Partenkirchen 1974, Congress Book, p. 511-515

[133] *B. Schwegmann,* “Wert und Unwert der heutigen Feinheitsmessung”, Farbe & Lack, Vol. 80, N° 10 (1974), p. 942-945

[134] *J. Kellner* and *F. Weyres,* “Methode zur photographischen Erfassung von Grindometeraufzügen”, Farbe & Lack, Vol. 77, N° 12 (1971), p. 1192-1196

[135] *DIN 53238-24:1995* Verfahren zur Beurteilung des Dispergierverhaltens. Teil 24: Mikroskopische Bewertung von Pigment und Füllstoffdispersionen

[136] *D. von Pigenot,* “Die Entwicklung der Farbstärke beim Anreiben von Pigmenten”, VIIth FATIPEC Congr., Vichy 1964, Congress B., p. 249-262

[137] *O. J. Schmitz, R. Kroker* and *P. Pluhar,* “Untersuchungen über die Dispergierbarkeit von Pigmenten” Farbe & Lack, Vol. 79, N° 8 (1973), p. 733-739

[138] *U. Zorll,* “Beitrag zur quantitativen Beschreibung der Farbstärkeentwicklung”, Farbe & Lack, Vol. 80, N° 1 (1974), p. 17-23

[139] *A. Klaeren* and *H. Völz,* “Versuche zur Farbstärkeentwicklung und quantitativen Beschreibung des Dispergiervorganges”, Farbe & Lack, Vol. 81, N° 8 (1975), p. 709-716

[140] *A. Lust,* “Online-Farbstärkebestimmung während der Dispergierung”, Farbe & Lack, Vol. 98, N° 10 (1992), p. 753-757

[141] *J. H. Braun,* “Gloss of Paint Films and the Mechanism of Pigment Involvement”, Journal of Coatings Technology, Vol. 63, N° 8 (1991), p. 43-51

[142] *J. H. Braun* and *D. Fields,* “Gloss of Paint Films: I. Effects of Pigment Size”, Journal of Coatings Technology, Vol. 66 N° 1 (1994), p. 93-98

[143] *H. D. Jefferies,* “The Dispersion of Titanium Dioxide Pigment and the Influence of the Degree of Dispersion on Some Film Properties”, J. Oil Col. Chem. Assoc., Vol 45, N° 10 (1962), p. 681

[144] *H. D. Jefferies* in G. Parfitt, “Dispersion of Powders in Liquids”, Applied Science Publishers, London and New Jersey, 3rd Edition, 1981, p. 460

[145] *P. Hauser, W. Jettmar* and *R. Sappok,* “Mikroskopische Untersuchung der Goniometamerie”, XIth FATIPEC Congress, Florenz 1972, Congress Book, p. 551-556

[146] *F. Finus,* “Messung des Helligkeitsflops von Metalleffektlacken mittels eines automatisierten Goniophotometers”, XIIth FATIPEC Congress, Garmisch-Partenkirchen 1974, Congress Book., p. 393-398

[147] *J. Oyarzún,* “Die Dispergierbarkeit von Pigmenten als Kriterium zur Ermittlung optimaler Mahlgutformulierungen”, XVIth FATIPEC Congress, Liège 1982, Congress Book, Vol. IV, p. 95-116

[148] *R. B. McKay,* “Rheological Behaviour and Colloid Chemical Charakter of Organic Pigment Dispersions”, XIIIth FATIPEC Congress, Cannes 1976, Congress Book, p.428-434

[149] *W. Herbst,* "The Dispersing of Pigments", reprint, originally published in German in Farbe & Lack, Vol. 76, N° 12 (1970), p. 1190-1208
[150] *W. Herbst,* "Untersuchungen über das Dispergierverhalten organischer Pigmente in Bindemitteln auf PVC-Basis". Deutsche Farben Zeitschrift, Vol. 26, N° 10 (1972), Part I, p. 519-532, Part II, N° 11, p. 571-576
[151] *W. Herbst,* "The Dispersion of Organic Pigments in Printing Inks and Paint Systems with Various Types of Dispersing Equipment", Progress in Organic Coatings (1972/1973) N° 1, p. 267-331
[152] *W. Herbst,* "Über die Dispergierung von Pigmenten mit modernen Dispergiergeräten" (1), Farbe & Lack, Vol. 77, N° 11 (1971), p. 1072-1090, and (2), N° 12, p. 1197-1203
[153] *D. P. Roelofsen,* "Entwicklungen auf dem Gebiet des Pigmentdispergierens mit Rührwerkskugelmühlen", Farbe & Lack, Vol. 97, N° 2 (1991), p. 235-242
[154] *J. Ruf,* "Organischer Metallschutz", Vincentz Verlag, Hannover, 1993, p. 573-580
[155] *K. Merkle* and *W. Herbst,* "Abhängigkeit der kritischen Pigmentvolumenkonzentration von den Dispergierbedingungen", Farbe & Lack, Vol. 77, N° 3 (1971), p. 214 -223
[156] *F. K. Daniel,* "Determination of Mill base Composition for High Speed Dispersers, Journal of Coatings Technology, Vol. 38, N° 500, (9/1966), p 535-542
[157] *O. Hafner,* "The influence of various parameters on the dispersibility of organic pigments in paint", J. Oil Col. Chem. Assoc., Vol. 57 (1974), p. 268-273
[158] *L. Dulog,* "Neue Theorien zum Dispergieren von Pigmenten und zur Dispersionsstabilität", XIXth FATIPEC Congress, Aachen 1988, Congress Book, Vol. I, p. 25-48
[159] *J. Winkler,* "Beschreibung des Dispergierprozeßes durch eine Wahrscheinlichkeitsgleichung", XIXth FATIPEC Congress, Aachen 1988, Congress Book, Vol. III, p. 289-306
[160] *J. Winkler, E. Klinke, M. Sathyanarayana* and *L. Dulog,* "Theory for the Deagglomeration of Pigment Clusters in Dispersion Machinery by Mechanical Forces, part II" Journal of Coatings Technology, Vol. 59, N° 754 (1987), S. 45-53
[161] *J. Winkler* and *L. Dulog,* "Theory for the Deagglomeration of Pigment Clusters in Dispersion Machinery by Mechanical Forces, part III", Journal of Coatings Technology, Vol. 59, N° 754 (1987), p. 55-60
[162] *J. Oyarzún,* "Ein neues Verfahren zur Ermittlung der Dispergierbarkeit von Pigmenten in mittelviskosen Medien", Farbe & Lack, Vol. 79, N° 6 (1973), p. 518-530
[163] *R. K. Eckhoff,* "Factors Affecting the Dispersibility of Powders", Chr. Michelsens Institut, Bergen, Vol. 39, N° 3 (1976), p. 1-20
[164] *J. Oyarzún,* "A new method for determining dispersibility of pigments and the optimal mill base formulation", J. Oil Col. Chem. Assoc., Vol. 61, N° 5 (1978), p. 160-168
[165] *F. K. Daniel,* "Getting the most out of a high-speed disperser", American Paints and Coatings Journal, N° 4 (1982), p. 60-64
[166] *H. Rohrer,* "Bestimmung der optimalen Mahlgutzusammensetzung zur Herstellung von titandioxidhaltigen Anstrich- und Druckfarben in verschiedenen Geräten", Farbe & Lack, Vol. 69, N° 8 (1963), p. 591-596
[167] *W. Göring* and *S. Paksoy,* "Studien zur optimalen Mahlgutformulierung von Pigmentpasten für die Sandmühle", Farbe & Lack, Vol. 79, N° 2 (1973), p. 99-107
[168] *W. Liehr,* "Ermittlung günstiger Mahlgutzusammensetzungen für verschiedene Anreibemaschinen", Farbe & Lack, Vol. 71, N° 8 (1965), p. 625-632
[169] *B. Schwegmann,* "Was ist und wie arbeitet ein Dissolver?", Deutsche Farben Zeitschrift, Vol. 20, N° 10 (1966), p. 478-480

[170]*anon.*, "Mill base formulations – dissolver", Bayer Technical Informations on the use of pigments in surface coatings, Farbenfabriken Bayer AG, 1992

[171]anon., "Tioxide Pigments in Non-Aqueous Media", Booklet BTP/162 of BTP Technical Service Department, British Titanium Products

[172]*S. Guggenheim,* "Data on Pigment Dispersion with High Speed Impeller Equipment", Official Digest, Vol. 30, N° 7 (1958), p. 729-735

[173]*J. Oyarzún,* "Dispersibility of Pigments as a Criterion for Determining Optimum Mill Base Formulations", Journal of Coatings Technology, Vol. 55, N° 704 (1983), p. 77-85

[174]*J. Oyarzún,* "Optimising Mill Bases by Statistical Experimental Design – L'optimisation des pâtes de broyage par conception expérimentale statistique", European Coatings Journal, copy 1-2/1994, p. 44-52

[175]*R. L. Plackett* and *J. P. Burman,* "The Design of Optimum Multifactorial Experiments" Biometrica, Vol. 33 (1946), p. 305-325

[176]*W, Carr* and *A. Kelly,* "Factors which Affect the Efficiency of Sand Grinding", J. Oil Col. Chem. Assoc., Vol. 62, N° 6 (1979), p. 183-198

[177]*L. Gall,* "Dispergieren von Pigmenten – Theorie und praktische Erfahrungen", Technical Information of Badische Anilin- und Soda-Fabrik AG, 1978

[178]*L. Gall* and *R. Siebke,* "Statistische Methoden bei der Optimierung von Mahlgutzusammensetzungen", IXth FATIPEC Congress, Brussels 1968, Congress B., p. 501-507

[179]*J. Winkler* and *L. Dulog,* "Dispergieren von Pigmenten in Lacken mit einer Rührwekskugelmühle", Farbe & Lack, Vol. 90, N° 4 (1984), p. 244-250

[180]*J. Winkler,* "Titanium Dioxide", 2nd Edition, Vincentz Network, 2013, p. 56

[181]*F. Hinze, S. Rippberger, M. Stintz,* Chem. Ing. Techn. 72, No.4, p. 322-332

[182]*R. W. O'Brien,* Journal of Fluid Mechanics 190 (1988) 71

[183]*R. W. O'Brien, L. R. White,* J. Chem. Soc. Faraday Trans. 2, 74 (1978) 1607

Index

A

absorption 66, 69, 70, 71, 73, 75, 77, 79, 81, 83
absorption coefficient 24, 66, 70, 71, 75, 79
acid-base character 134
acid-base concept 137
adhesion force 14, 126
adhesive tension 42
adsorption isotherm 30
adsorption isotherm according to *Brunauer*, *Emmett* and *Teller* 32
agglomerate 14, 119, 177, 178, 180
agglomerate, breakdown 120
aggregates 14
anchor group 131, 134
anisotropy 16
anomalous dispersion 71
anomalous flow 91
anomalous flow property 188
anti-floating agent 163
anti-flooding agent 163
anti-settling agent 115
Antonov, G. N. 44
Antonov's rule 44
Apel, K. 35, 110
apparent viscosity 94
Arrhenius, S. 88
assessment of transparency 77
attractive energy 138
average equivalent diameter 34
average particle diameter 20
average particle size 34, 73, 122

B

Balfour, J. G. 153
Bayer 166
Bénard cell 157, 161, 163
BET method 29, 35
binder concentration 188
binder demand 197, 198, 204
binder shock 165
Bingham body 96
Bingham viscosity 96
Bingham yield value 96
Bingham, E. C. 96
Boltzmann distribution 145
Boltzmann, L. 145
bonding strength 133
Bragg, W. H. and *W. L.* 17, 18
Brandts-Buys, L. 137
Braun, J. H. 54, 181
bridging flocculation 130
Brockes, A. 82
Broglie, L. de 22
Brownian motion 24, 130, 139, 144
Brunauer, S. 29, 32
Burman, J. P. 211
Burrell, H. 136, 137

C

capillarity 45
capillary pressure 46, 128
Carr, W. 82, 135, 212
Casson relationship 97
Casson viscosity 98
Casson yield value 98
Casson, N. 97, 110
Chapman 144
chemisorption 29
CIE 76
Clayfield, E. J. 141
coarseness 169, 172, 174
coefficient (scattering and absorption) 75
coflocculation 155, 163
cohesion forces 120, 122, 123, 126
cohesive forces 43
Colour Index 14
competitive adsorption 135
concentration (of pigment) 109
concentration triangle 204
contact angle 41, 42, 127

Coulombic force 14, 19, 122, 131, 164
Coulombic repulsion 143
Coulter counter 27
critical pigment volume concentration 187
critical surface tension 43, 127
crystal modification 15
crystal structure 15, 17, 19

D

Daniel, F. K. 203
Debye, P. P. 47, 121
Debye-Scherrer 17, 19
Debye-Scherrer method 17
degree of dispersion, assessing 169
degree of efficiency 206, 210
degree of flocculation 153
dependence of shade on particle size 82
depletion stabilisation 141
depth of shade 79, 150, 161, 175
Deryaguin, B. V. 143
determination (of scattering and absorption coefficients) 75
determination of hiding power 76
dichroism 16
dielectric constant 146, 147, 149
diffuse double layer 144
diffuse reflection 54
dilatancy 94, 112
dipolar force 29
dispersibility 21, 28, 50, 176
dispersing equipment 186
dispersion (of white light) 71
dispersion curve 71, 171
dispersion force 137
dispersion hardness 123
dispersion process 169
Dittrich, H. 135
Dörfler, H.-D. 41
doughnut effect 205
Dulog, L. 188, 83
Dupré, A. 126
dynamic light scattering 24

E

ease of dispersion 124
effective volumen 189
effective wetting 135
Einstein, A. 154
Einstein-von Smoluchowski, equation 154
Eissler, R. L. 137
electric double layer 120, 143
electric double layer, electric surface charge 120
electrical polarisability 47
electroacoustic method 27
electron microscopy 22
electrophoretic mobility 147, 148
electrostatic forces 96
electrostatic repulsion 130, 143, 146
electrostatic stabilisation 143, 147, 162
Emmett 32
Emmett, P. H. 29
energetic efficiency factor 190
energy barrier 138, 155
energy density 190, 192
enthalpic stabilisation 142
enthalpic/entropic stabilisation 142
entropic elasticity 106
entropic stabilisation 142
equivalent diameter 20, 27
extinction 65
extinction coefficient 66

F

Feigin, R. I. 141
fictitious yield value 98
Fields, D. 181
final tinting strength , 175, 176, 179
final viscosity 97, 100
fineness of grind 169, 172, 202
Finus, F. 183
floating 150, 156
flocculate 15, 113
flocculation 50, 117, 130, 142, 146, 159, 161
flocculation number 161
flocculation rate 153
flooding 150, 156
flop 182
flotation 156
flow 96
flow anomalies 109
flow behaviour 107, 108

flow curve 91
flow property 50, 111
flow property of plastic melt 117
flow property, printing ink 111
flow-point curve 203
flow-point method 204
flow-point procedure 203
fluidity 87
four-colour printing 73
Fraunhofer evaluation method 26
free surface energy 39
Fresnel formulae 55
Fresnel, A. 57
Freundlich adsorption isotherm 29
Freundlich, H. 30

G

Gall, L. 214
Gans, R. 74
Garmsen method 171
Garmsen, W. 170
Gauss, C. F. 46
Gaussian distribution 29
Gauss-Laplace equation 46
gelation 116
gel-curve 102, 103
Gibbs, J. W. 141
gloss 54, 55, 57, 115, 150, 169, 153
gloss development 181
gloss haze 54, 55, 57, 59, 60
gloss measurement 58, 182
Goetze scale 174
Goetze, W. 174
goniochromatism 182
Göring, W. 204
Gouy, G. 144
Gouy-Chapman 144
grain hardness 123
grindometer 170, 184
Guggenheim factor 205
Guggenheim, S. 205

H

Hadjistamov, D. 99
Hafner, O. 83
Hagen, G. 128
Hagen-Poiseuille's law 128
half-life value 195, 198
Hamaker constant 140
Hamaker, H. C. 140
Hamann, K. 133
hardness scale 123
Haug, R. 166
Hauser, P. 82
Hegman device 184
Hegman system 170
Hegmann 170
Helmholtz, H. von 141, 143, 144
Herbst, W. 83
Hergesell, E. 135
Herschel-Bulkley 98, 101
hiding power 16, 61, 69, 70, 74, 78, 115, 119, 150, 169
Hildebrand, J. 136
Hird, M. J. 153
Hoeflack, M. 116
Honigmann, B. 111
Hooke, R. 104
Hooke's law 104
hue 81, 119
hue flop 182
Hulst, van de 63
Huyskens, P. 137
hydrogen bond 29, 96, 131
hydrogen bonding 122, 133
hydrogen bonding force 43, 136, 137
hydrophilicity 37, 39
hydrophobicity 38

I

increase in tinting strength 176
influence of the solvent 135
initial and final tinting strength 176
initial tinting strength 179
inorganic pigment 70, 73
interfacial tension 40
interference structure 68
isoelectric point 148
isomorphism 16, 20

J

Jefferies, H. D. 182

Joppien, G. 133

K

Kaluza, U. 161
Kämpf, G. 82
Kaufmann, W. 82
Keesom, W. H. 121
Keifer, S. 153
Kellner, J. 173
Kelly, A. 212, 214
kinematic viscosity 87
Klaeren, A. 180
Klinke, E. 188
Kresse, P. 163
Kubelka-Munk relationship 161, 174
Kubelka-Munk theory 75, 80, 81

L

Lambert, J. H. 65
Lambert's law 65
Landau, L. D. 143
Langmuir adsorption isotherm 31, 132
Langmuir, I. 30, 33
Laray 118
laser Doppler anemometry 148
laser light diffraction 24, 25
Laue, M. von 17, 19
let-down 164, 165
Liehr, W. 167
lightening power 78
lightness flop 182, 183
London constant 139
London force 121, 122, 131
London or dispersion force 29
London, F. 121
Lumb, E. C. 141
Lust, A. 180
lyophilic 126
lyophobic 126

M

Maikowski, M. A. 79, 82
master-batch 117
Maxwell body 106
Maxwell, J. C. 105
Maxwell's equation 55
Maxwell's field equation 64
mean rate of dispersion 176, 179
measuring, flocculation behaviour 151
mechanical barrier 138
mechanical breakdown 120
metallic-effect coating 182
method of *Debye* and *Scherrer* 18
Mie effect 63, 65, 73
Mie scattering 63, 69, 72
Mie scattering parameter 61
Mie theory 25, 61, 63, 65, 67, 72, 73, 74, 82
Mie, G. 61, 65, 67, 69
mill base optimisation 203, 206, 209
mill base preparation 186
mill base viscosity 185, 189
mixed-phase pigment 16, 20
mobility (of pigment particles) 141, 152, 157, 159, 161
Mocker apparatus 22
Mohs 123
Molenaar, F. 116

N

Napper, D. H. 132, 133, 141, 142
Newton, I. 87
Newton's law of viscosity 105
Newtonian fluid 22, 108, 114
non-linear plastic 111
non-linear plastic substances 100
non-linear plasticity 97

O

Oesterle, K. M. 135
oil absorption value 206
optimum particle size 68
optoelectronic grindometer 171
organic pigment 70, 73
Ostwald equation 93, 98
Ostwald, W. 93, 98
Overbeek 143, 146
Oyarzún, J. M. 198

P

Pahlke, H. 113

Paksoy, S. 204
particle diameter, average 20
particle morphology 13
particle shape 16, 74, 109
particle size 79, 111, 169
particle size and morphology 22
particle size and scattering 72
particle size and shape 69
particle size distribution 21, 27, 28, 82, 109
particle size optimum 68
particles, primary 14
Patton, T. S. 117, 118, 124
photosedimentometer technique 23
physisorption 29
Pierce, P. E. 109, 114
Pigenot, D. von 175, 180
pigment blend 162
pigment concentration 201, 207
pigment dispersibility 188, 193, 206
pigment preparation 117
pigment shock 164
pigment volume concentration 74, 100, 158, 182
pigment, binder and solvent concentration 185
Plackett, R. L. 211
plastic 114
plastic-dilatant flow 100
pleochroism 16
polarisability 37, 48, 62
polarity 37, 39
polydispersity index 25
polymer adsorption 130, 131, 135, 141
polymorphism 15
pour test 152
prevention of flocculation 156
primary particles 119, 177
printing ink, flow property 111
pseudoplastic 114, 188
pseudo-wetting 135

R

Rayleigh scattering 61, 62, 69, 72
Rayleigh, W. S. 61, 68, 72
readily dispersible pigment 50
re-agglomeration 120, 130, 166, 180
reflectance 75
reflection 54, 55, 57, 58
reflectivity 55, 56
refraction 60
refractive index 16, 48, 69
refractive index and absorption 70
Řeháček, K. 134
representative viscosity 94
repulsive energy 138
rheogram 91
rheology 21, 85
rheopexy 103
Roelofsen, D. P. 187
Rohrer, H. 204
Rothstein, E. C. 134
rub-out number 161
rub-out test 152, 160, 174
Ruf, J. 187
Rumpf, H. 125

S

Sander, H. 166
Sappok, R. 38
Sato, T. 134
saturation 78, 81
scattered infrared radiation 153
scattering 60, 61, 69, 71, 73, 75, 77, 79, 80, 81, 83
scattering coefficient 66, 70, 75, 153
scattering cross sections 66
scattering efficiency 66
Schaeffer, C. 119
Schäfer, W. D. 134
Schmidbauer, H. 102
Schmitz, O. J. 177, 180
Schwegmann, B. 172
sedimentation test 151
seeding 166
semi-transparency 70, 77, 150
shade 82
shear rate 87, 109
shear stress 86, 112
shock effect 164
shortness 113
Siebke, R. 214

sieve analysis 21
silicone oil 163
Smoluchowski, M. von 153, 154
sol-curve 102, 103
solubility parameter 136
solvation 135, 137
solvent adsorption 131, 135
solvent shock 166
Sørensen, P. 134, 137
specific enthalpy of wetting 38
specific surface area 20, 29, 109, 110, 115
specular reflection 54
sphericity value 34
spray-pouring test 160
spreading 42
stabilisation 120, 130, 149
standard depth of shade 81
statistical design 208
steric hindrance 130, 138
Stern, O. 144
Stokes, G. G. 24
Stokes' law 22, 24, 151, 154, 158
Stokes-Einstein equation 24
structural viscosity 92
sum distribution 28
surface character 37, 109
surface charge 156, 162
surface energy 126
surface tension 40, 129, 157, 162, 165, 166
surface treatment 51, 134
surface-treated pigments 211
surfactant 162, 163, 186

T

tack 113
Teller, E. 29, 32
testing, flocculation behaviour 151
theory of Mie 24, 79
thickening 116, 162
thickness of the adsorbed layer 134
thixotropic loop 101
thixotropy 101, 117, 150
tinting strength 50, 51, 169, 202, 206, 209, 212
tinting strength development 174
total interaction energy 138
transparency 50, 51, 77
transparency number 77

U

ultrasonic attenuation spectroscopy 26

V

van der Waals attraction forces 29
van der Waals force 14, 15, 43
velocity gradient 87
Verwey, E. J. W. 143, 146
viscoelasticity 112
viscosity 125, 129, 130, 137, 160, 192, 215
viscosity coefficient 91
viscosity curve 91
viscosity function 93
Vogel 90
Vold, M. J. 141
Völz, H. 80, 180

W

Waals, van der 96
Washburn equation 129
Washburn, E. 128
water-thinnable binder system 107
Weber, H. H. 73
Weissenberg effect 106
wet-point 204, 205
wetting 38, 40, 43, 47, 120, 125, 129, 113, 162
wetting energy 125
wetting kinetics 128
wetting tension 42, 127
wetting-out 186, 192
Weyres, F. 173
Winkler, J. 188, 164
work of adhesion 126

X

X-ray crystallography 17

Y

yield value 95, 100, 111, 112, 114, 117, 150

Young, T. 44
Young's equation 42, 44, 127

Z

zeta potential 145, 147
Zettlemoyer, A. C. 37, 112, 113
Zisman, W. A. 43
Zorll, U. 54, 117, 180

www.ingramcontent.com/pod-product-compliance
Lightning Source LLC
LaVergne TN
LVHW021725021026
849161LV00004BA/727